Free Enterprise Environmentalism

Free Enterprise Environmentalism

Walter E. Block

LEXINGTON BOOKS
Lanham • Boulder • New York • London

Published by Lexington Books
An imprint of The Rowman & Littlefield Publishing Group, Inc.
4501 Forbes Boulevard, Suite 200, Lanham, Maryland 20706
www.rowman.com

6 Tinworth Street, London SE11 5AL, United Kingdom

British Library Cataloguing in Publication Information Available

Library of Congress Cataloging-in-Publication Data

Names: Block, Walter, 1941- author.
Title: Free enterprise environmentalism / Walter A. Block.
Description: Lanham : Lexington Books, 2021. | Includes bibliographical references and index. | Summary: "For all too long, 'environmentalism' has been a wholly-owned subsidiary of the political left. This book, however, makes the case that markets, free enterprise, limited government, private property rights are a far better way of addressing ecological challenges than are our present institutions of socialism, government control and regulations"—Provided by publisher.
Identifiers: LCCN 2021039007 (print) | LCCN 2021039008 (ebook) | ISBN 9781498586856 (cloth) | ISBN 9781498586863 (epub)
Subjects: LCSH: Environmental economics. | Environmentalism—Economic aspects. | Free enterprise.
Classification: LCC HC79.E5 B59 2021 (print) | LCC HC79.E5 (ebook) | DDC 338.9/27—dc23
LC record available at https://lccn.loc.gov/2021039007
LC ebook record available at https://lccn.loc.gov/2021039008

To all our children, and to their children, and to their children's children.

Contents

Preface

In the view of the "watermelons" (green on the outside, red on the inside) free enterprise environmentalism is a contradiction in terms. If a person favors the one, he necessarily opposes the other. This is because, in my view, such people have an indomitable desire to control the lives of others; they are merely looking for a new means to achieve this end, now that communism has been shown unable to do so.

In very sharp contrast, the present book takes the position that markets, private property rights, and free enterprise are the last best hope for saving the environment; that problems such as air pollution, species extinction, deforestation, and water shortages are the results of government intervention into the economy, failure to uphold private property rights, and state ownership and mismanagement.

Most people, I fear, have been taken in by this siren song of socialist environmentalism. Nowadays, everyone who is anyone is ostentatiously recycling, supporting alternative energy sources (but *not* nuclear power), trying to leave a smaller carbon "footprint," extolling the virtues of Al Gore, and so on. This includes pundits, editorialists, sociologists, grade and high school teachers, literary people, clergy, historians, feminists, political scientists, biologists, chemists, and other economically illiterate individuals. It even includes some economists who really should know better.

The book that follows this preface consists of a series of chapters I have published over the years. How did I first get interested in this field? It all started with my daughter, Hannah. When she was about seven or eight years old, she came home one day from school and said something to the effect of "Daddy, you're not acting to save the planet." I knew full well from where she got this viewpoint, from her leftie teacher in the second or third grade. I was more than a bit annoyed at this "environmentally correct" brainwashing of small children on the part of public school teachers. I was ready to "pounce" on my daughter and try to explain my own side of this issue. But my wife ordered me not to do so on the ground that if I convinced Hannah of free-market environmentalism, and she started articulating these views in class, she would be a pariah among her school fellows and teachers. Did I really want to do that? No, I suppose, I did not. She was way too young to take on such a burden. So, instead, gentle reader, I am instead, now, in effect, taking it out on you if you continue reading. Fair is fair. You are adults; my daughter was not, at least at that time.

Another reason for writing this book is to save the word "environmentalist" for the good guys. Right now, to be an "environmentalist" is to favor government ownership of property, statist regulations of everything from air pollution to the amount of water in a toilet bowl, and more and more "public" encroachments on the private sector. But "environmentalism" is an approbation. Why should we stand idly by and allow the enemies of a civilized order to steal it for their own deleterious purposes? If there is such a thing as a "free enterprise environmentalist" then there cannot, logically, be any plain old ordinary "environmentalists." No, opponents, if the philosophy espoused in this volume, must then be more accurately characterized as they are: socialist environmentalists, or interventionist environmentalists, or some such. Not only will this effort, if successful, contribute to a more accurate nomenclature, it may, to some small degree, turn us onto a more ethical and benevolent path in the future.

Walter Block
New Orleans, Louisiana, 2021

Section I

INTRODUCTION

Chapter 1

Private Property Rights and Environmentalism

INTRODUCTION

This chapter shall attempt to reconcile environmentalism and economic freedom.[1] Before making this seemingly quixotic endeavor, we must be sure we are clear on both concepts. Environmentalism may be non-controversially defined as a philosophy which sees great benefit in clean air and water, and to a lowered rate of species extinction. Environmentalists are particularly concerned with the survival and enhancement of endangered species such as trees, elephants, rhinos, and whales, and with noise and dust pollution, oil spills, greenhouse effects, and the dissipation of the ozone layer. Note that this version of environmentalism is a very moderate one. Moreover, it is purely goal directed. It implies no means to this ends whatsoever. In this perspective, environmentalism is, in principle, as much compatible with free enterprise as it is with its polar opposite, centralized governmental command and control.

Economic freedom also admits of a straightforward definition. It is the idea that people legitimately own themselves and the property they "capture" from nature by homesteading,[2] as well as the additional property they attain, further, by trading either their labor or their legitimately owned possessions.[3] Sometimes called libertarianism, in this view the only improper human activity is the initiation of threat or force against another or his property. This, too, is the only legitimate reason for law. The only proper functions of legal enactments are the prevention of murder, theft, rape, trespass, fraud, arson, and so on, and all other such invasions.

At first glance the relationship between environmentalism and freedom would appear direct and straightforward: an increase in the one leads to a decrease in the other, and vice versa. And, indeed, there is strong evidence for an inverse relationship between the two.

For example, there is the Marxist and even communist background of some advocates of environmental concerns.[4] People like these come to the ecological movement with an axe to grind. Their real interest is with power: running the lives of others, whether for their own good, for the good of society, or for the good of the unstoppable "forces of history." They were doing pretty well on this score for decades in Russia and Eastern Europe. Thanks to them, this vast part of the globe was marching in lockstep toward the Marxist vision of all power to the "proletarians." But then, in 1989, thanks to the inner contradictions of communism (Mises, 1969), their world turned topsy turvy. Some shifted their allegiances to the only fully communist systems remaining: Cuba, North Korea. As for the others, nothing daunted, they just switched horses on the same old wagon: instead of formal socialism, these people adopted environmentalism as a better means toward their unchanged ends. They can best be characterized as "watermelons" in that while they are green on the outside they are still red on the inside.

Then, there are the real greens. They see environmentalism not as a means toward an end, but as the very goal itself. The most radical of them is very forthright. They see man as the enemy of nature, and would, if they could, destroy the former to save the latter. Graber (1989, p. 9), who is a U.S. National Park Service research biologist, states: "Until such time as Homo Sapiens should decide to rejoin nature, some of us can only hope for the right virus to come along." In the view of Foreman (1990, p. 48), who is co-founder of Earth First![5] and former lobbyist for the Wilderness Society, "We are cancer on nature." And here is how Mills (1929, p. 106) describes the other member of her own species: "Debased human protoplasm."[6]

Some are only slightly less radical. They do not yearn for virtually the end of the human race. Instead, they merely hold that animals have rights, that trees have rights, that microscopic organisms have rights. It is reputed that Gandhi, for instance, sometimes went around wearing a surgical mask, so that he would not inadvertently kill a microorganism by inhaling it. If so, that practice would certainly be in keeping with this philosophy.

Stepping down a peg in the extremism of the ecologically concerned, there are those who merely blame the markets, free enterprise, and capitalism, for the ruination of the planet. In their view what is needed is to curb these vicious appetites, and to return to a "kinder, gentler" version of governmental interventionism. For example, with regard to contaminated New York City beaches, the Commissioner of Health from the Big Apple stated on Canadian Public Television (July 30, 1988):

> I think the motivation is greed, you know non-caring about the planet, non-caring about the ocean, and not caring about the people who live on the planet and want to use the ocean—greed.

In the view of environmentalist Renate Kroisa regarding pulp mill operators (CTV Report, March 15, 1989):

> They would rather rape the environment and make a lot of money for themselves than not rape the environment, clean up, and later on . . . stay competitive. The mills are here to make a lot of profit, and they are making a lot of profit at the cost of our environment.[7]

And states Commoner,[8]

> The origin of the environmental crisis can be traced back to the capitalist precept that the choice of production technology is to be governed solely by private interest in profit maximization.

Other Statements of this ilk include Porrit and Winner (1928, p. 11): "The danger lies not in the odd maverick polluting factory, industry or technology but in . . . industrialism itself" Bookchin (1970, p. 14): "The plundering of the human spirit by the marketplace is paralleled by the plundering of the earth by capital"; and free markets "take the sacredness out of life, because there can be nothing sacred in something that has a price" (Schumacher, 1973, p. 45).[9]

Then there are those who oppose not only market competition in general but also want to ban particular products made possible by this system. For instance, there are calls to prohibit 747 airplanes (Rifkin, 1980, p. 216), automobiles (Sale, 1989, p. 33), eyeglasses (Mills, 1989, p. 106) private washing machines (Bookchin, 1989, p. 22), tailored clothing (Schumacher, 1973, pp. 37–58), toilet paper (Mills, 1989, pp. 167–162).

Paradoxically, there is a very limited but possible sense in which it is rational to prefer the reds to the greens. True, the former, not the latter, killed millions upon millions of people (Conquest, 1986, 1990). But at least their goal, their purpose, their aim, their end, was to help human beings. Yes, they picked a tragically erroneous way of going about this, a philosophy from which the entire world's peoples are still reeling. However, it must be conceded that they were not traitors to their species.[10] This, unfortunately, cannot be said of some of the greens, particularly the more radical ones. Nor can it be denied that at least so far, with the exception of a few unfortunate loggers, the greens have not killed nor hurt very many people. But if their own publicly articulated intentions are to be believed, given the power they might be a greater danger to the human race than even the communists.

This, in short, is the case for believing there to be an inverse relationship between environmentalism and freedom. However, it is not a direct and straightforward one: an increase in the one does not always lead to a decrease in the other, and vice versa.

What are the exceptions? How can environmentalism and economic freedom be reconciled?[11] Simple. By showing that free enterprise is the best means toward the end of environmental protection. This appears a daunting task at the outset, given the emphasis placed by most environmentalists on socialism, and their hatred for capitalism. But a hint of the solution may be garnered by fact that laissez faire capitalism, as adumbrated above, strenuously opposes invasions, or border crossings, and that many environmental tragedies, from air pollution to oil spills, may be reasonably interpreted in just such a manner. The reason for environmental damage, then, is the failure of government to protect property rights (omissions) and other state activity which either regulates private property, or which forbids it outright (commissions). Let us consider a few test cases.

AIR POLLUTION

According to the mainstream economic analysis, libertarianism is wrong. The problem of airborne pollution is not due to a failure of government to protect private property rights. Instead, this comes about because of "market failure," a basic flaw in free enterprise. Pigou (1912, p. 139) gives the classic statement of this view:

> Smoke in large towns, which inflicts a heavy loss on the community, . . . comes about because there is no way no to force private polluters to bear the social cost of their operations.

Samuelson (1936, 1970) conveys the same sentiment in terms of the divergence between private and social costs. Lange and Taylor (1938, p. 103) are yet additional socialists who make the complementary point:

> A feature, which distinguishes a socialist economy from one based on private enterprise, is the comprehensiveness of the items entering into the socialist price system.

In other words, for some strange dark mysterious reasons, capitalists, under laissez faire, are excused from even considering the physical harm they do to the property of others through the emissions of their smoke particles. Under socialism, in contrast, the central planner of course takes this into account, nipping the problem of pollution in the bud.

There is so much wrong with this scenario it is hard to know where to begin a refutation. Perhaps we may best start with an empirical observation. If this criticism of the market were true, one would expect that, even if the Soviets couldn't successfully run an economy, they could at least be trusted as far as the environment is concerned. In actual point of fact, nothing could be further from the truth.

Exhibit "A" is perhaps the disappearance of the Aral and Caspian Seas, due to massive and unchecked pollution, overcutting of trees, and consequent desertification. Then there is Chernobyl, which caused hundreds, if not thousands of deaths.[12] For ferryboats in the Volga River, it is forbidden to smoke cigarettes. This is not for intrusive paternalistic health reasons as in the West, but because this river is so polluted with oil and other flammable materials that there is a great fear that if a cigarette is tossed overboard, it will set the entire body of water on fire.

Furthermore, under communism, there was little or no waste treatment of sewage in Poland, the gold roof in Cracow's Sigismund Chapel dissolved due to acid rain, there was a dark brown haze over much of East Germany, and the sulfur dioxide concentration in Czechoslovakia were eight times levels common in the United States (DiLorenzo, 1990A).

Nor was it a matter merely of the absence of democracy in the USSR. The ecological record of the U.S. government, where democracy is the order of the day, is none too savory. The Department of Defense has dumped 400,000 tons of hazardous waste, more than the five largest chemical corporations combined. The Rocky Mountain Arsenal carelessly disposed of nerve gas, mustard shells, the anti-crop spray TX, and incendiary devices. And this is to say nothing of the infamous Yellowstone Park forest fire, which the authorities refused to put out, citing ecological considerations;[13] nor the TVA's 59 coal-fired power plants; nor the underpricing and overuse of land administered by the Bureau of Land Management; nor the fact that the government subsidizes forest overcutting by building logging roads.

These are not examples of market failure. Rather, they are instances of government failure: direct controls and inability or unwillingness to uphold private property rights.

But what of Pigou and Samuelson's charge of the misallocative effect of negative externalities, or external diseconomies? This, too, is erroneous.

Up to the 1820s and 1930s, the legal jurisprudence in Great Britain and the United States was more or less predicated upon the libertarian vision of non-invasiveness (Coase, 1960; Horwitz, 1977).

Typically, a farmer would complain that a railroad engine had emitted sparks which set ablaze his haystacks or other crops. Or a woman would accuse a factory of sending airborne pollutants to her property, which would dirty her clean laundry hanging on a clothesline. Or someone would object to the foreign matter imposed in one's lungs without permission. Almost invariably, the courts would take cognizance of this violation of plaintiff's rights.[14]

The usual result during this epoch was injunctive relief, plus an award of damages.

Contrary to Pigou and Samuelson, manufacturers, foundries, railroads, and so on could not act in a vacuum, as if the costs they imposed on others were of no moment. There was a "way to force private polluters to bear the social cost of their operations": sue them, make them pay for their past transgressions, and get a court order prohibiting them from such invasions in future.

Upholding property rights in this manner had several salutary effects. First of all, there was an incentive to use clean-burning, but slightly more expensive anthracite coal rather than the cheaper but dirtier high-sulfur-content variety; less risk of lawsuits. Second, it paid to install scrubbers and other techniques for reducing pollution output. Third, there was an impetus to engage in research and development of new and better methods for the internalization of externalities: keeping one's pollutants to oneself. Fourth, there was a movement toward the use of better chimneys and other smoke prevention devices. Fifth, an incipient forensic pollution industry was in the process of being developed.[15] Sixth, the locational decisions of manufacturing firms were intimately affected. The law implied that it would be more profitable to establish a plant in an area with very few people, or none at all; setting up shop in a residential area, for example, would subject the firm to debilitating lawsuits.[16]

But then, in the 1840s and 1850s, a new legal philosophy took hold. No longer were private property rights upheld. Now, there was an even more important consideration: the public good. And of what did the public good consist in this new dispensation? The growth and progress of the U.S. economy! Toward this end it was decided that the jurisprudence of the 1820s and 1830s was a needless indulgence. Accordingly, when an environmental plaintiff came to court under this new system, he was given short shrift. He was told, in effect, that of course his private property rights were being violated; but that this was entirely proper, since there is something even more important than selfish, individualistic property rights. And this was the "public good" of encouraging manufacturing.[17]

Under this legal convention, all the economic incentives of the previous regime were turned around 180 degrees. Why use clean-burning, but slightly more expensive, anthracite coal rather than the cheaper but dirtier high-sulfur-content variety? Why install scrubbers, and other techniques for reducing pollution output, or engage in environmental research and development, or use better chimneys and other smoke prevention devices, or make locational decisions so as to negatively impact as few people as possible? Needless to say, the incipient forensic pollution industry was rendered stillborn.

And what of the "green" manufacturer, who didn't want to foul the planet's atmosphere, or the libertarian, who refused to do this on the grounds that it was an unjustified invasion of other people's property? There is a name for such people, and it is called "bankrupt."[18] For to engage in environmentally sound business practices under a legal regime which no longer requires this is to impose on oneself a competitive disadvantage. Other things equal, this will guarantee bankruptcy.

From roughly 1850 to 1970, firms were able to pollute without penalty. *This* is why "there is no way to force private polluters to bear the social cost of their operations" à la Pigou; *this* is why there was a Samuelsonian "divergence of social and private costs." This was no failure of the market. It was a failure of the government to uphold free enterprise with a legal system protective of private property rights.

In the 1970s a "discovery" was made: the air quality was dangerous to human beings and other living creatures. Having caused the problem itself, the government now set out to cure it, with a whole host of regulations which only made things worse. There were demands for electric cars, for minimal mileage per gallon for gasoline, for subsidies to wind, water, solar, and nuclear power,[19] for taxes on coal, oil, gas, and other such fuels, for arbitrary cutbacks in the amount of pollutants into the air. The nationwide 55 miles per hour speed limit was not initially motivated by safety considerations, but rather by ecological ones. "Rent seeking" played a role in the scramble, as Eastern (dirty-burning sulfur) coal interests prevailed over their Western (clean-burning anthracite) counterparts. The former wanted compulsory scrubbers; the latter wanted the mandated substitution of their own coal for that of their competitors.

And what was the view of the supposedly free market–oriented Chicago School? Instead of harking back to a system of private property rights, they urged the "more efficient" statist regulations. Instead of a command and control system, they urged the adoption of tradable emissions rights (TERs). In this system (Hahn, 1989; Hahn and Stavins, 1990; Hahn and Hester, 1989), instead of forcing each and every polluter to cut back by, say, one-third, they would demand of all of them together that this goal be attained. Why is this beneficial? It might be difficult and expensive for some firms to reduce pollution from 150 to 100 tons, and easy and cheap for others. Under TERs, some could reduce the pollution levels by less than one-third (or even increase them), while they would, in effect, pay others to reduce theirs by more than this amount. The means through which this would be accomplished would be a system of "rights to pollute," and an organized market through which these could be bought and sold.

The implications of this scheme for freedom are clear. States Anderson (1990):

Fortunately, there is a simple, effective approach available—long appreciated but under used. An approach based solidly on . . . private property rights.

At its root all pollution is garbage disposal in one form or another. The essence of the problem is that our laws and the administration of justice have not kept up with the refuse produced by the exploding growth of industry; technology and science.

If you took a bag of garbage and dropped it on your neighbor's lawn, we all know what would happen. Your neighbor would call the police and you would soon find out that the disposal of your garbage is your responsibility, and that it must be done in a way that does not violate anyone else's property rights.

But if you took that same bag of garbage and burned it in a backyard incinerator, letting the sooty ash drift over the neighborhood, the problem gets more complicated. The violation of property rights is clear, but protecting them is more difficult. And when the garbage is invisible to the naked eye, as much air and water pollution is, the problem often seems insurmountable.

We have tried many remedies in the past. We have tried to dissuade polluters with fines, with government programs whereby all pay to clean up the garbage produced by the few, with a myriad of detailed regulations to control the degree of pollution. Now some even seriously propose that we should have economic incentives, to charge polluters a fee for polluting—and the more they pollute the more they pay. But that is just like taxing burglars as an economic incentive to deter people from stealing your property, and just as unconscionable.

The only effective way to eliminate serious pollution is to treat it exactly for what it is—garbage. Just as one does not have the right to drop a bag of garbage on his neighbor's lawn, so does one not have the right to place any garbage in the air or the water or the earth, if it in anyway violates the property rights of others.

What we need are tougher clearer environmental laws that are enforced—not with economic incentives—but with jail terms.

What the strict application of the idea of private property rights will do is to increase the cost of garbage disposal. That increased cost will be reflected in a higher cost for the products and services that resulted from the process that produced the garbage. And that is how it should be. Much of the cost of disposing of waste material is already incorporated in the price of the goods and services produced. All of it should be. Then only those who benefit from the garbage made will pay for its disposal."[20]

Economic freedom thus implies a movement back to the legal status of pollution in the earlier epoch. Nor need we fear undue economic hardship and dislocation because of adjustment problems. For apart from obvious and blatant pollution, which has already been curtailed through command and control regulations, it will take at least a few years for environmental forensics to develop to the point where industry will have to make more basic changes.

There are, of course, objections to "turning the clock back" to the 1820s. For one thing, there is the fear that if we allow anyone to sue anyone else for pollution, that will mean the end of industry altogether. And not only of industry and other accoutrements of modern civilized life. This would also bring the curtain down on life itself, as, strictly speaking, even exhaling (carbon dioxide) could be seen as a pollutant, and thus forbidden. Fortunately, this scenario is not tenable. First of all, although industry up to the 1830s was no great shakes compared to the modern era, it was not as non-existent as implied by this objection either. Second, there is a reason for this: the burden of proof is on the plaintiff, so only the more egregious cases of pollution were, in effect, actionable, and de minimis was in operation, so that frivolous lawsuits or ones alleging only tiny amounts of pollution were disregarded.[21]

Another objection, a more reasonable one, is that if allowing pollution lawsuits again will not bring industry to a screeching halt, it will at least greatly disorganize it. Perhaps it might be better to allow for a 10-year waiting or warning period, so that industry could adjust before imposing so draconian a set of measures.

This option does indeed sound more pragmatic, but there are problems with it. We have said that pollution amounts to an invasion. Suppose that someone had the authority to immediately end an invasion, say, for example, slavery, and refused to do so for 10 years on the grounds that this would be too "disruptive" or "impractical." Say what you will about such a decision on pragmatic grounds, it cannot be maintained that it enhances freedom.

Fortunately, we can have our cake and eat it too in the present context. That is, we can allow environmental lawsuits immediately, but also have a "waiting period" of perhaps 10 years or so in any case. This can be accomplished because of the 150-year gap, from approximately 1843 to 1993, when environmental forensics could have developed, but did not, thanks to a legal regime which was not conducive to it.[22] The point is, had environmental forensics been developing over these last 150 years, but for some reason not implemented, and we were to suddenly allow environmental lawsuits for the first time at present, this would indeed drive industry to an abrupt halt. For the plaintiff's burden of proof would be easy to satisfy; under these assumptions. Moreover, there would be plenty of invasive pollution around to find people guilty of perpetrating.

For with emissions strictly controlled (in the early period), development would have proceeded along non-pollution-intensive lines. In contrast, with carte blanche on emissions (the later period), industry would have developed in a pollution-intensive manner. Moving from a system where pollution was all but legal (1843–1970), to one where it was strictly controlled (as it was before 1843), would thus have called for a basic restructuring of industry.

Let me try to make this point in another way. There is a difficulty which the private property rights theory of environmental protection must wrestle with: if we institute such a system abruptly, especially if we did so, say, in the 1960s before these concerns had captured the public imagination, we ran the risk of halting industry dead in its tracks, something to be resisted at the very least on pragmatic grounds. On the other hand, if we offered, for example, a ten-year waiting period before environmental lawsuits could be undertaken, then we are complicit in violations of the libertarian code during this decade. Happily we can avoid this dilemma. First, we allow lawsuits as soon as we have the power to do so, thus escaping from the second (disrupting industry) horn of the dilemma. We escape from the first, too, because of the fact that for the plaintiff to be successful in his lawsuit he must prove beyond a reasonable doubt that a specific, particular polluter is responsible for invading his person or property. But to do so, given the sad sorry state of environmental forensics at least at the time of this writing, will take time, plausibly, as much time as it will take for industry to end the error of its ways without any great disruption. That is, suppose it takes 40 years for industry to adjust to the legal dispensation of the 1830s. This will not be as harmful to the economy as might be supposed because it might take a similar amount of time to figure out precisely who is polluting whom.[23]

WASTE DISPOSAL

The brou-ha-ha over paper vs. plastic and Styrofoam wrappers also has implications for economic freedom.

In the late 1980s, a McDonald's restaurant opened its doors in Moscow. In some ways, this was no great shakes. Ray Kroc's burger emporia had by that time been doing business in hundreds of other countries. But in other ways, this was a very big deal indeed. For at that time Russia was still under the control of communism. Allowing a private firm to do business in the heart of the beast thus showed a weakness in the totalitarianism of the USSR. What could be a greater chink in the armor than a popular restaurant intimately tied to Western capitalism?

McDonald's is a reasonable example of a capitalist enterprise. It employs thousands of people, particularly young persons, minority members, and immigrants. It brings joy to millions of customers, and has sold, almost unbelievably, in the billions of burgers. It is an indication of quality. You can travel practically the world over, and be assured of the same kind of meal they serve in "Kansas." This chain (and other imitators) has been a boon to the poor. Before its birth, it was difficult for the poor to enjoy a restaurant meal; thanks to this initiative, away from home dining has become a commonplace for those with modest means. All in all, McDonald's was not a poor choice as a chip in the high stakes gambling with the communists over the future of the world's political economy.

But at about the same time that Ronald McDonald was taking up residence behind the Iron Curtain, back at home in "the land of the Free and the home of the brave," he was running into entry restrictions and other barriers. Dozens of town councils, all across this great land of ours, were refusing to give McDonald's permission to open up new stores. Why? Is it a takeover of Soviet fifth columnists, or communist revolution, in the good old U.S. of A? Not a bit of it. Instead, it was all due to left-wing environmentalism.

Why were the local greens so bitterly opposed to more quarter pounder outlets? Because they came wrapped in Styrofoam and other plastic packaging, and if there is one thing practically guaranteed to drive an environmentalist to apoplexy, it is precisely these materials.

Let us assume, merely for the sake of argument, that everything any ecologist has ever said about plastic and Styrofoam is true. That compared to paper, these substances are not environmentally friendly, they are not biodegradable, they are not recyclable, they are not reusable, and they cannot be returned to nature. On the contrary, when buried in the ground, they come back to haunt us in the future as hazardous wastes. And that as a result, anyone foolish enough to dispose of them ruins his land for subsequent farming, housing, factories, shopping malls, and so on.

Under these conditions, let us enquire into the ability of the marketplace to transmit this knowledge (paper, good; plastic, bad) so that it is taken into account by the economy. After all, this is precisely what the price system is presumably designed to do. Prices, after all, are like street signs. Just as the latter guide us around geographical space,[24] the former is supposed to impose direction on the economy.

At first glance it would appear that while prices might accomplish their task in the general economy, they are a dismal failure when it comes to environmental concerns. Picture yourself at the supermarket checkout counter. You have just selected your groceries, and the clerk has charged you for them. After paying, you are asked that inevitable, fateful $64,000 question: paper or plastic bag?

Under these circumstances, the only reason for picking the environmentally sound paper, and eschewing the toxic plastic, is benevolence. For let us assume that the cost to you is $0.01 for each. In some cases, this is explicit. You pay a penny for either one. In other cases, it is only implicit: you don't pay for the bag, paper, or plastic; rather, it is included in the price of the groceries, in much the same manner as the lighting, or cleaning, or advertising of the store. Benevolence for the planet, or for your fellow creatures is your only possible motivation for choosing the paper; for by stipulation the economic considerations are equal—one penny for each.

But we all know what Smith (1776) said about benevolence. It is not from benevolence, but rather from a keen appreciation of self-interest, that the butcher, the baker, and the candlestick maker share with us their wares. Given, the evils of plastic, benevolence is a weak reed indeed upon which to base our hopes for its elimination. Nor is it a question of benevolence versus selfishness. Given the importance of ridding the planet of these noxious substances, it behooves us to mobilize both motivations, not just one of them. Benevolence is far from sufficient. For suppose that half of all industrialists had the personality of a libertarian Mother Teresa, and refused to pollute, even though allowed to do so by 1aw. What would happen to them? They would go bankrupt, for they would give themselves a competitive disadvantage. If all industrialists are roughly of equal ability, but some pollute and others spend money on smoke prevention devices, it is clear that the invisible hand will be choking us, not helping us. No, the only solution is to change the law to one which upholds property rights so that trespassers do not continue to be privileged.

Why has the price system seemingly failed? Is this intrinsic to capitalism, one of the "market failures" that socialist economists are always prattling on about? Not at all! The failure stems not from laissez faire, but from state prohibition. Specifically, the government has nationalized, or municipalized, the industry of solid waste management.

Right now, we do not pay a red cent for garbage disposal. Instead, we are forced by government to disburse tax money for this purpose, and are then given these services for "free." In other words, this service is run along the lines of socialized medicine. There, too, services are provided for "free," courtesy of our tax dollars.

These systems have several disadvantages.[25] For one thing, there is the phenomenon of "moral hazard." Charge people a very low or zero prices, and they will buy much more than at normal prices. Further, they will "waste" the good or service.[26] This is seen in the fact that socialized medicine is a hypochondriac's dream come true and that consumers purchase items which are promiscuously wrapped. Given that the housewife doesn't have to pay to dispose of package coverings, it is no wonder that the manufacturer has little incentive to economize on containers.[27] How would a private market in garbage disposal function? Everything would be privatized—The trucks which make pickups from the homeowner as well as the dump-sites themselves. There would be no mandatory recycling or bottle deposit requirements;[28] there would only be laws against trespassing: disgorging waste material onto other people's private property. How would prices be established? Assume that burying inoffensive paper costs only a penny per bag, but that the plastic variety is so harmful that each one does $5.00 worth of damage[29] to the land in which it is placed. Given competition, no dumpsite owner will be able to charge more than $5.00 for burying a plastic bag, lest the additional profits earned thereby attract new entrants into the industry. In like manner, the price cannot fall below this amount, since if it does, it will bankrupt all who agree to it. For example, if a private dumpsite owner were to accept $4.00 compensation for agreeing to permanently store a plastic bag on his land, he would lose $1.00 on that transaction. Multiply this by a few truckloads of this substance, and he will no longer be able to continue in business.[30]

Now let us return to our supermarket checkout scenario. Only this time, under full privatization, we make an entirely different economic calculation than before.

Previously, there was no impetus to choose either paper or plastic. Each cost $0.01, and that was the end of it. Now, however, matters are very different. For we are called upon not merely to purchase the bag material, but also to dispose of it later on at our own expense. Given disposal costs of one penny for paper and five dollars for plastic, our total costs are readily calculated: two cents for paper, and five dollars and one cent for plastic.

Is there any doubt that the whole problem would disappear in one fell swoop under these economic conditions? Virtually no consumer in his right mind would choose environmentally unfriendly plastic. The costs would simply be prohibitive. Everyone would "do the right environmental thing" and select paper.

This does not mean, of course, that plastic bags would be totally banned by economics. They would still be utilized, but only when their value to the user was greater than $5.01. For example, blood, intravenous solution, and other medical fluids might still employ plastic containers.

Thus, thanks to the "magic of the market," we can again have our cake and eat it too. Under a full private property rights regime, there is no reason to legislatively ban McDonald's. If plastic and Styrofoam are truly hazardous to the health of the planet, they will impose tremendous costs on dumpsite owners. These will be passed on to consumers. If McDonald's continued to insist upon use of plastic and Styrofoam, this firm would lose out to other competitors (Burger King, Wendy's, Pizza Hut, Taco Bell, A&W, etc.) who were more greatly concerned for their customers' pocketbooks. Under present assumptions, there is simply no need to reduce freedom in order to protect the planet. The two work in tandem.

Table 1.1

	Purchase Cost ($)	*Disposal Cost ($)*	*Total Cost ($)*
Paper	0.01	0.01	0.02
Plastic	0.01	5.00	5.01

Source: Data based on author calculations.

But it is now time to question our assumptions about the relative harm to the planet of plastic and Styrofoam. According to "garbologist"[31] Rathje (1989), plastic is not so much a hazard to the globe as it is inert. If there is anything dangerous to the planet it is paper; not in the form of bags, but rather telephone books. After many years of burial, these release methane gas, and other dangerous substances. If so, perhaps paper and plastic will be able to compete with one another on a somewhat more equal footing.

This is an empirical question, which cannot be decided on the basis of armchair economic theorizing. It may safely be left in the hands of the private dumpsite owning industry, for these entrepreneurs, unlike environmental bureaucrats, stand to lose their own personal fortunes if their prices are not consonant with actual harm to their property, and hence to the environment in general.

CONCLUSION

I have tried to show that in at least two cases, air pollution and waste disposal, the concerns of environmentalists and those who favor economic freedom can be reconciled. However, there might appear to be what one anonymous referee called a "basic structural flaw in the development of (my) argument" in that the public policy conclusions in each of these two cases appear to be very different. "On the one hand," continues this referee, "[I] applaud . . . the existence of a pre-1850 legal system which enforced private property rights. But in (my) final argument for letting the market control waste disposal there is no clear indication of what if any role environmental law would play."

I am tremendously grateful to this referee in that he has given me an opportunity to further explicate libertarian environmentalist theory. The seeming contradiction in how I handle the two ecological issues can be reconciled in this way. In the case of air pollution, the violation of economic freedom and private property rights was that polluters were allowed by law to in effect trespass on other people's land, to say nothing of their lungs. In the case of waste disposal, the breach of economic freedom and private property rights is no less apparent, although it takes an altogether different form. Here the infraction consists of the nationalization (e.g., municipalization) of what would otherwise be private dumpsites. But in both cases there is a transgression of the free enterprise ethic. Therefore, in each, the capitalist-oriented environmentalist will advocate a return to market principles. In the one case this consists of an end to legal trespass, in the other of privatization of garbage dumping. Thus, there is no "structural flaw," or indeed, any inconsistency whatever, in this analysis.

Let me make this point in a different way. Egalitarian socialists oppose both income disparities and private medicine. For the former they advocate wealth redistribution; for the latter, socialized medicine. Now these are two very distinct things. Seemingly, they are incompatible with one another. But not really, since both are aspects of an underlying vision.

It is the same in the present case. Laws prohibiting trespass of smoke particles, and privatizing dumpsites are superficially very different. In actual point of fact they are but opposite sides of the same coin, in that they both emanate from the same philosophical principle.

One last point! The typical way of treating pollution in the literature is as an "externality." By now it should be clear that I totally reject this approach. An external diseconomy is defined as a harm perpetrated by A on B, one for which B can neither collect damages nor halt through injunction. But why is B so powerless? It is my contention that the victim of pollution finds himself in this precarious position solely because of inadequacies in the law. Previous to 1850, for example, there was no pollution externality. This came about due to a "government failure" to uphold the law against trespass, not because of any alleged, "market failure" such as externalities.

NOTES

1. The author wishes to thank Jonathan Adler of CEI, Dianna Reinhart, Jane Shaw, and Rick Stroup of PERC, P.J. Hill of Wheaton College and Jan Leek of NCPA for bibliographical and other help. None are responsible for the content of this chapter. The author wishes also to thank two anonymous referees for very substantive help in the rewriting of this chapter. Without their assistance this article would have been a far less cogent one.

2. For a critique of homesteading, see Stroup (1988). For a rejoinder, see Block (l990B): for another defense of homesteading, see Hoppe (1993).

3. For a general explication of the private property-based free enterprise system; see Rothbard (1973), Hoppe (1989). For political economic perspectives that are sometimes confused with this vision, see Hayek (1973), Nozick (1974). For a rebuttal of these, see Rothbard (1982B).

4. Names that come to mind in this regard include Tom Hayden, Jane Fonda, Helen Caldicott, Jeremy Rifkin, Kirkpatrick Sale, and E. F Schumacher. For discussions of this phenomenon, see Horowitz (1991), Bramwell (1989), Rubin (1994) and Kaufman (1994).

5. This is the group that urges tree spiking; placing a metal spike in trees so that when the chain saw of the lumberman encounters it, his injury or even death will result. Their rallying cry slogan is "Back to the Pleistocene."

6. These views were cited in Goodman, Stroup et al. (1999, p. 3).

7. Reported in "On Balance," Vol. II, No. 9, 1989, p. 5.

8. Cited in DiLorenzo (1990A).

9. Cited in Goodman, Stroup et al. (1991, p. 4).

10. It is on this ground that the communists may be preferred to the Nazis. For apart from members of the Aryan nations, the Nazis actually did intend to, and actually succeed in, killing massive numbers of people. In terms of actual numbers of people killed, however, the reverse is the case.

11. For a book that attempts to do just this, see Block (1990A).

12. True, there is the U.S. counterpart nuclear meltdown at Three Mile Island. But a popular bumper sticker puts this into some sort of perspective. It stated: "More people died at Chappaquidick than at Three Mile Island." ("Chappaquidick" refers to the death of a single individual, Mary Jo Kopechne, while being driven by Senator Ted Kennedy.) The point is, of course, that no one, not a single solitary individual, lost his life at Three Mile Island.

13. Forest fires, it turns out, are "natural," and nothing must be done, which interferes with nature.

14. Called at the time "nuisance suits," we can with hindsight see them as complaints.

15. It is only because murder and rape were illegal that there was a call for a forensic industry, capable of determining guilt based on semen, blood, hair follicles, DNA, and so on. If these activities were legal, these capabilities would not have developed. Similarly, when one can sue for pollution, it is of the utmost importance to determine guilt or innocence; hence, the establishment of environmental forensics.

16. Of course, "coming to the nuisance" was not deemed acceptable. That is, one could not build a residential abode in an area first homesteaded by pollution emitters, and then sue for pollution. On this see Rothbard (1982A, 1990).

17. As a sop to the plaintiffs, the law and judicial practice was altered so as to require very high minimum heights for smokestacks. In this way the local perpetrator of invasive pollution no longer negatively impacted the local plaintiff. But of course this did no more than sweep the problem under the rug, or, rather, into the clouds. For if polluter A no longer affected complainant A, he affected others. And polluters B, C, D who previously did not harm A, now began to do so.

18. This is the exact opposite of Adam Smith's (1776) "invisible hand." Ordinarily, in laissez faire capitalism, selfish seeking of profit leads to the public good. For example, one invests in a good, which is in very short supply and hence most needed by the populace, and earns the greatest possible profit. Here, instead, if a person acts in an environmentally responsible manner, he goes broke.

19. The Price Anderson Act—protecting firms from legal responsibility for accidents—is the most egregious case of the latter.

20. For another critique of tradable emissions rights, see McGee and Block (1994).

21. On this see Rothbard (1982A, 1990).

22. From 1845 to 1970, approximately, polluters had a free run of the atmosphere, other people's property and their lungs. From roughly 1970 to 1995, and counting, there was concern for invasive air and water-borne pollutants, but only command and control (and in the last few years, tradable emissions rights schemes) regulations. Provision for environmental lawsuits is still, as of this 1995 writing, virtually nonexistent. See Horwitz (1977), Block (1990, pp. 282–285).

23. I am grateful to an anonymous referee for forcing me to clarify my presentation of this point.

24. In the days of yore when a city was faced with a besieging army, one of the defensive measures they would take was to tear down the street signs. They did so on the ground that this would hardly much dis-accommodate long-time citizens, but would play havoc with the invader's ability to travel around town.

25. Apart, that is, from the immorality of forcing people, whether by democratic vote or no (Spooner, 1870, 1966), to pay for things they have no desire to purchase.

26. If we ran a socialized milk program as we do garbage disposal and medical care, people would probably have "milk gun" (instead of water gun) fights, wash their cars with milk, and take milk baths.

27. In addition to excessive amounts of wrappings, our zero price policy has also led to the combination of different materials in them, such as paper, plastic, tin and other metals, cardboard, and so on. All of this makes it more expensive to recycle.

28. Which are but further infringements upon economic freedom.

29. Science cannot at present precisely determine the amount of damage that might be caused. (I owe this point to an anonymous referee.) However, this presents no philosophical challenge to entrepreneurship. Those dumpsite owners whose predictions are the closest to reality will prosper, at least compared to their colleagues furthest away, given ceteris paribus conditions. But make no mistake about it; given the assumptions on the basis of which we are now operating, storing paper most certainly will harm the dumpsite itself, at the very least in terms of economics. For, to reiterate, we are presuming that buried plastic and Styrofoam have much the same effect as a toxic waste. Those dumpsite owners who allow storage of these items under their land will reduce its economic value after the landfill is complete, and alternative uses (housing, farms, etc.) are contemplated.

30. I am here implicitly assuming that the present discounted (dis)value of burying a single plastic bag is $3.00. Obviously, to charge only $4.00 for this service would be to lose money on the deal.

31. A garbologist is to mounds of waste material as is an archeologist to ancient ruins. Each yearns to "get to the bottom" of their respective subject matters. Each analyses them from their own perspective.

Section II

WATER

Chapter 2

Ocean Real Estate

INTRODUCTION

In October 2006 the journal *Nature* published an article on the effects of commercial fishing on fish stock variability, prompting news stories in the following days warning about the risk to some fish species. The conclusion of this *Nature* article is that the commercial fishing effects are more complex and profound than had been previously thought, leading to a potential collapse and extinction of some species (Hsieh, 2006). Bruce Lieberman quotes one author of the study saying that "the only practical solution may be to establish more marine reserves, which ban or limit fishing and other sea harvesting to protect various species" (Lieberman, 2006). This response is typical of those who wish to protect the environment by simply restricting access to it.

For thousands of years, free access to ocean fisheries was the norm and human impact from fishing was negligible. However, modern commercial fishing equipment and techniques and the increased numbers of fishermen have produced a sizable reduction in many fish populations. The ocean has been a common resource for all to exploit and the result is the well-recognized tragedy of the commons. However, the solution to this tragedy is not to impose government regulations and restrictions, but to privatize the ocean in order to more efficiently manage it as a valuable natural resource and also open the possibility to radically increase its availability and productivity.

Michael Markels, CEO of Ocean Farming, Inc., reported in 1995 plans to run trials of large-scale ocean fertilizing projects based on new research and technology. Ocean fertilizing could increase the nutrients available to promote growth of the food chain for many commercially and environmentally valuable fish species. Undertaking investment projects of this sort could dramatically enhance the productivity of ocean fisheries, but ocean fertilizing is expensive and risky, and under the current circumstances, others could benefit from the investments without the expense (Markels, 1995). Increasing the fish stock in the ocean is technically feasible, but the lack of property rights may be the biggest obstacle.

BACKGROUND

Currently, devastating environmental conditions exist in the ocean because no one owns the property and, therefore, no one has a stake in what happens to this resource that houses an array of valuable assets—fish, oil, minerals, transportation, recreation, and so on. Governments cause this wreckage by not recognizing private ownership of this wealth, except the few miles they own off the coast for military protection.[1] If private property were established, it would improve the environmental problems that occur in the ocean such as endangered species, wetlands devastation, and pollution.[2]

In the ocean, many fish that are near extinction are specifically the ones seafood eaters love the most, like Atlantic cod, Atlantic and Pacific salmon, marlin, swordfish, snappers, tuna, sharks, and Alaskan king crabs (DeMont, 2003). No one owns the ocean or the commodities in it, so the incentive, which is similar to the incentive for forest devastation on government land, is for people to fish as much as possible. Therefore, others should not be amazed that these fish are on the verge of extinction, because they are free-floating assets. The same result would happen if there were gold nuggets dispersed in the local public parks. Eventually, people will collect virtually every piece.

History serves as a lesson, which is seen in the near extinction of the buffalo. No one owned this animal and, so it fell victim to a tragedy of the commons due to excessive hunting.[3] On the other hand, the cow that was privately owned and existed at the same time as the buffalo was not subjected to overhunting and has lasted until this day. This is because cow owners invest money in buying, feeding, and protecting their cows. Therefore, when one of their cows is killed, the owner faces a cost: loss of the animal in the future. But if the owner kills a buffalo, he faces no such cost. Both these animals are of value for their meat and hides, but, consequently, only one was almost wiped out because of the lack of private ownership.

As modern fishing techniques emerged, the pressure on fisheries began to increase. Fishermen noticed declining catches and began to have greater and greater difficulty finding schools of fish. These conditions also lead to conflict among fishermen blaming each other for their mutually incompatible actions. The "tragedy of the commons" was underway at sea.

In response to declining fish stocks, governments began the process of regulating access to the fisheries by command and control. The primary regulatory measure implemented was to limit the Total Allowable Catch (TAC) (Runolfsson, 1997). This type of regulation is still in place in many countries with the predicable result of a race to harvest that catch and bring it to market.

Mandel (2006) complained that the national regulatory agencies are too "clumsy" to efficiently enforce their limits. Berkes (2006, p. 1557) refers to those who are overfishing around the world as "pirates" and "roving bandits." Their research suggests that these activities could likely lead to mass ocean fish extinctions. They properly identify free access to the fisheries as the problem, but labeling the fishermen as "roving bandits" implies that they are stealing someone else's property. These authors concede that the establishment of property rights would create incentives for more efficient fish harvests, yet they call for more regulatory efforts to limit deliveries to international fish markets as an important part of the solution (Berkes, 2006, p. 1558). Regardless of the regulations, the real problem is fishermen are heavily depleting unowned stocks in unowned waters. If they are bandits, from whom are they stealing?

PARTIAL RIGHTS

More recently, an innovative solution has been the establishment of "partial" property rights to various fish stocks. The government imposes fishing quotas, which would limit the amount of fish captured and save the stock, by requiring permits for individual transferable quotas (ITQs), which acts as partial private property (Whitehead and Block, 2002).[4] ITQs do not work as optimally as private ownership of fish, but at least give the permit holders a stake in the supply of fish (Whitehead et al., 2004). Because the fishermen purchased the ITQ permits through auction, they will lose some or all of their investment if the fish supply dwindles. The ITQ permit holders' investment loss, however, is much smaller than it would be to private fish owners if the supply dwindled. It is one thing to suffer a decrease in the value of what is in effect only a hunting license. It is quite another to lose the entire property. Thus, private ownership is a better solution for that one reason alone, because there is a bigger stake in the fish supply and incentive to maintain it. Nevertheless, ITQ owners because of their best interest are likely to try and preserve the fish stock than without the permits (EU, 2005).

There is another problem with ITQs. They are a version of market socialism. They are the aquatic counterpart to the economic system once employed in Tito's Yugoslavia.[5] In both cases, there is no pure private property. Rather, the government owns the resources. The governments "owned" all factors of production in Eastern Europe. In the case of the ITQs, the governments "own" the fish. It cannot be denied that there is a "market" aspect of the "market socialism" in these cases. But this resemblance is by no means to be conflated with true markets, based on private property rights. In ITQs, for example, the fishermen may only bid for the right to *capture* the fish. They do not own the fish themselves.

In order to preserve the fish stock, the ITQ owners will have to lobby the government (EU, 2005). After the establishment of the ITQs, the fishermen and environmentalists are united on the policy against overfishing and both would like to see the species' lives protected, which is unlike the status quo ante ITQs when both were on opposing sides of a zero-sum game. Despite both parties being in favor of species preservation, this does not mean the government will protect the fish. If the fish supply dwindles or is eliminated, the government, which would be responsible for conserving the valuable resources, would not suffer any losses or investment if it did nothing to improve the fish count. The government does not own the fish nor is it a fishing company, which would give it the incentive to protect these resources. Rather, just like any other institution, it comprises self-interested individuals, except government officials have limited, fixed terms that make them unaccountable for the actions they take, especially with taxpayers' money. Therefore, environmental solutions should not be left to the whims of greedy politicians.

PRIVATE PROPERTY

Although ITQs lessen the environmental problems more than without them, they are only a meager substitute for the optimal solution, which is private property. Any private animal farm is proof that privately owned animals are far from extinction. Some private farms can actually restock the wild when certain animals become too scarce, which is what happened with alligators in Louisiana (Mayell, 2001). The alligator in Louisiana is a good example of how private property solved the tragedy of the commons. The alligator was close to extinction, because people were overhunting them primarily for their skins but also their meat. A law was passed in 1963 to ban hunting (Mayell, 2001). In Australia, people also were overhunting crocodiles, but private crocodile farms solved the problem.

In 1972, biologists in Louisiana proposed a "Marsh to Market Plan," which requires private ownership of alligators to protect them from poachers (Mayell, 2001). Privately owning alligators has boosted the population substantially, and in 1987 the government took alligators off the endangered species list. In fact, "About 75 percent of all wild alligator hides, along with about 85 percent of all farmed skins, used by tanners around the world come from Louisiana"

(Mayell, 2001). Alligator privatization has also helped to preserve Louisiana wetland property, which is beneficial for hurricane buffering and serves as the habitat for many animals. Unfortunately, many people are not aware that the alligators are no longer endangered, and the resulting decreased demand for them has reduced the incentive for current producers to better manage a growing population of alligators. On the contrary, environmentalists who do want to help the wetlands should consider purchasing alligator products such as purses, wallets, boots, and belts; of course, an alligator po-boy would help (Robichaux, 1989).[6] When people purchase more of these alligator skin items, it increases the demand and price for the skins, which gives farmers a bigger incentive to maintain or increase the alligator population.

Environmentalists are also worried about the devastation of wetlands through offshore oil drilling. Again, these environmental problems can be solved through cooperative negotiations between private property owners, which is seen best between the Audubon Nature Society, and an oil company. The Audubon Nature Society is a private environmental organization, which owns land all over the United States and uses it to protect wildlife and other habitats. This Society opposes development on government land by private corporations. The Audubon Nature Society, however, realized that the cost of not allowing oil companies to drill on its property in Louisiana and Michigan was too high (Anderson and Leal, 2001, pp. 84–85). The organization decided to allow the drilling on the land under strict stipulations, in order to minimize the damage done to the environment and wildlife. Because of the limitations the society imposed, it reduced the amount of profit that could have been realized. Audubon, however, valued the environment much more than the expected royalties that could have otherwise been made from allowing complete environmental devastation.

The income derived from oil in Louisiana allowed the Audubon Society to purchase other land to further conserve the environment. Therefore, private property forces each party to realize the costs associated with its actions, but the government, a third party, interferes in the market process when it claims the property.[7] Anti-market environmentalists pay plenty of attention to offshore drilling, including the effects of oil pollution in the water. Oil pollution can be broken down as follows: "60% routine tanker transport operation, 20% oil spills, 15% natural oil seepage from the seafloor and sediment erosion, and 2.5% from platforms" (Lomborg, 2001, p. 189, 393). But, pollution is a violation of people's property rights. Courts must uphold the rights of individuals to not have pollutants trespass on their property (Clanton, 2006). If someone pollutes anything on someone else's property, the polluter should be held accountable. Additionally, if property owners allow pollution on their property, and it spreads to another's, then the property-owning polluter should be held accountable (Rothbard, 1982).

Therefore, ocean owners would have a strict liability that does not now exist, because private property in the ocean does not exist. Since 60 percent of oil spills in the ocean is caused by routine tanker transportation, ocean owners would be more careful not to allow leaky tankers through their property for two reasons. First, pollution devalues property, and no one will want the losses to their property by letting others pollute freely. Second, they are liable if the oil in their property trespasses to others' property. One way to uphold property rights is to hire ocean-keepers who would inhibit trespassing.[8] They would assess payments on tankers based on the risk of liability they add for traveling through the property. They would likely charge significantly less for environmental-friendly tankers with double hulls.

Additionally, if any damage occurs to the property that is caused by oil drilling, then the culprits will have to compensate the property owners. Currently, pollution and environmental damage exist in excess of an ideal amount, because no one owns the property to realize these costs of pollution. The government has restrictions and regulations for ocean pollution, as it does for both overfishing and wetland devastation. The government though is limited to the amount of ocean it can control, because no one or country in the world claims ownership of it. Water has in effect become the tragedy of the global commons. Similar to the ITQs, the government issues tradable emissions rights (TERs) to allow people to pollute into the air. Besides the federal laws heavily restricting pollution in the ocean, the government would propose to solve ocean pollution through tradable ooze rights (TORs).[9] Again, this solution is secondary when compared with the complete benefits that private property offers.

Other methods of establishing property rights have been suggested for non-migratory fisheries. These rights are referred to as exclusive user rights (EURs) and territorial user rights in fisheries (TURFs) (Runolfsson, 1997). Although migratory species and the fluidity of the ocean itself presents challenges, the full establishment of property rights using some territorial boundaries is necessary and becomes even more feasible with modern technologies such as GPS monitoring. Of course, fisheries are not the only concern in exploiting the resources of the ocean. Full privatization of the ocean must be able to deal with mineral rights, pollution trespass, and large-scale ocean aquaculture. Perhaps precedents established by these user rights can create a foundation for defining other rights for understanding the development of ocean property.

HOW TO PRIVATIZE

Privatizing the ocean sounds like a great idea, but how is it possible? How can people distinguish masses of water from each other? The solution is quite simple. Currently, there is no demand for a technique to demarcate ocean property, because the idea of ocean real estate does not exist. If people, however, knew that ocean real estate would one day be feasible, then products to demarcate water property would be created. By responding to incentives in the market, entrepreneurs would produce different products that would satisfy consumers' wants for ocean property demarcation, in exchange for money. These product solutions could be electronic fences made specifically for water or GPS systems. Once people are given their rights to trade water, there will be a known demand for a product that demarcates water property. A solution will be invented. If our society can land a man on the moon, cure dreaded diseases, with the right economic incentives, it can certainly invent aquatic barbed wire, and even move from hunting to herding fish.

Besides demarcation, trespassing can be identified through other forms of technology that are not yet fabricated. Similarly, home security systems used to not exist, but the product was introduced after there was a feasible demand. Some people are very strict about trespassing and want no one else on their property. This does not necessarily mean that they will not allow ocean animals like fish to swim through their property, which is similar to the scenarios where animals roam across different private properties such as deer, rabbits, raccoons, or squirrels. People do not mind animals wandering around on their property. Although the animals they do not like, property owners will prevent them from trespassing, like crows in a corn field or rats in a house.

Entrepreneurs operating in markets are geniuses when it comes to finding solutions to satisfy people's wants. The problems associated with private fish farms in the middle of the ocean will also be solved with incentives in the private market to create technology. Traditionally, cattle were branded and dogs wore collars on their necks to identify owners. Today, people can implant RFID chips in their dogs to track them, if they are lost (Waters, 2006).[10] These chips could be a solution to branding the free-roaming private fish stock or using a laser device to emboss the owner's symbol of ownership into the skin of the fish. While chips are expensive today, a larger demand will create an incentive to expand supply and produce better technology to reduce the production costs, which can even now be seen through the decreasing prices of computers. Furthermore, if owners do not want their fish to leave their property, then other solutions can be found through profit-driven entrepreneurs. In the past, entrepreneurs responded to market demand by providing invisible electric animal fences that send electric shock through animals when they cross the boundaries marked off by the invisible fence. The same sort of remedy can be used for ocean animals.

A critical aspect of allowing markets to function efficiently is to make permanent commitments to the formation of property rights. Many countries have established tradable quotas, but they seem to only exist as a privilege by the government which might be revoked at any time. Of course, a revocable privilege does not alter the dynamics of overfishing much at all because a long-term perspective or investment can still not be established.

New Zealand has established quotas which are transferable, divisible, and permanent. Their more advanced property rights development has allowed for the creation of additional market-developed institutions which can further advance the cause of better resource management in fisheries. One such market development is the formation of a quota management corporation in which the owners of quotas can deal with their shared interests and resolve disputes by contract (Runolfsson, 1997). Ideally, they should have the right to fully manage the fisheries, but they have nevertheless organized in a way to promote voluntary cooperation.

CONCLUSION

The UN's Law of the Sea Treaty attempts to perpetuate the concept of the ocean as a "Common Heritage of Mankind." Under this concept the opponents of privatization continue to promote the ocean as a common resource. But to reduce the impact of the tragedy of the commons that would naturally flow from such a policy, they prefer the use of zoning laws for international waters and the establishment of international authorities such as the International Seabed Authority to dictate policy for the regulation and use of the ocean's resources. This concept is actually a perversion of property rights in that government and its agencies supposedly act on behalf of true owners which is all mankind. Full privatization must be based, rather, on individual rights of users, as it has long been so successfully done on land.

The idea of privatizing the ocean sounds absurd when you first hear it. Once you get used to the idea, then water becomes analogous to quick moving land. People centuries ago could never imagine individually owning land, because the king owned all of it. Yet, it is now commonplace and rightfully so. People make the best use of their land, because they have invested their resources in it. They will also make the best use of their water. When the government regulates the market, it inhibits the economy from optimal performance. Woefully, the U.S. government owns 271 million acres of land, which decreases the amount of agreements that could be made by private interest groups (Anderson and Leal, 2001). Water comprises 75 percent of the earth, and we have privatized less than the remaining 25 percent of land. With private property put to its most efficient use, the privatization of water will surely lead to the same positive results. As there will be no complete deforestation of land with private property, there will not be the ocean equivalent with fish extinction and over stripping of the ocean's resources.

As one can see, the market is full of potential solutions to problems that people of different interests face. Through private property, all are able to realize the costs of conserving their resources. These private property solutions apply to any problem as they do to overfishing, wetland devastation, and water pollution. People debating environmental issues should sincerely consider this alternative, instead of the mainstream's solution of more and more government intervention. The government already intervenes into our lives enough and has not solved the problem yet; giving them more power will not solve this matter, either. The private alligator farm proved the best solution to the dwindling alligator supply when compared to its government alternative of banning alligator hunting, which just led to poaching. Private property allows for the most efficient use of the earth's resources. People only own a small percentage of the world, imagine if more private ownership were allowed. It would surely prove beneficial for mankind.

NOTES

1. Historically, coastal nations defended a 3-mile limit under the concept of freedom of the seas. The UN Law of the Sea Treaty adopted in 1982 established a 12-mile territorial limit and a 200-mile Exclusive Economic Zone for coastal nations (www.heritage.org/Research/InternationalOrganizations/wm470.cfm). Beyond the 200-mile limit are international waters.
2. Swamps have been banished through linguistic legerdemain, so we need not worry about them.
3. Free access or use of a good by the public results in misuse and greater scarcity of that good.
4. There are a finite amount of permits and whoever bids the most for the permits will be allowed to fish.
5. Other versions of market socialism besides ITQs include tradable emissions rights for pollution (McGee and Block, 1994), and school vouchers (Rothbard, 1995; Vance, 2004).
6. A po-boy is a type of sandwich, which is made of French bread and meat in Louisiana that is similar to New York subs.
7. This cost refers to the opportunity cost, which represents the value of something you forgo by doing something else.
8. Ocean-keepers are the equivalent of gatekeepers that will allow entry into the private property.
9. Tradable ooze rights are the ocean equivalent of air's tradable emissions rights.
10. Radio Frequency Identification.

Section III

ENERGY

Chapter 3

Nuclear Power

INTRODUCTION

Nuclear energy has been heavily subsidized, and regulated, from its inception. The unintended consequence of these policies, however, was a distortion in the price of nuclear power, masking true costs of the investment, and result in gross inefficiencies. We propose a free market solution for this market. Subsidies, via the Price-Anderson Act, and regulations via the Nuclear Regulatory Commission (NRC), should be removed. In fact, we advocate the removal of all governmental distortions in the energy market.

First, we will discuss the history of the Price-Anderson Act, and how it distorts market signals. Then we turn our attention to the NRC's regulations—how these stifle innovation, and increase investment uncertainty. We find that the NRC's regulatory philosophy is inherently flawed and argue instead for free market "regulation" via insurance companies. Finally, we discuss how a capitalist system would address the problems of regulation, nuclear waste disposal, and the NIMBY problem.

Should there even be a nuclear industry? Our position is neither for nor against. We advocate a hands-off policy where this initiative is left to its own devices, free from governmental regulations and subsidies: free to succeed or fail on its own.

The Price-Anderson Act

The U.S. government initially became involved in the development of nuclear power plants so as not to be outstripped by the Communist Bloc in the kilowatt race (Morone and Woodhouse, 1989, p. 51). That is to say, the U.S. government undertook to beat eastern planning with western planning. The problem was that semi-private enterprise[1] was not developing and producing viable forms of nuclear power fast enough, at least by the standards of elected officials. One issue involved the liability insurance regarding catastrophic atomic accidents. Because private insurers would only provide coverage up to $50 million and it was believed that the actual cost of such an event would be much more, Congress passed the Price-Anderson Act in 1957. By enacting this law, the government limited the amount of damages to be paid in the event of a nuclear catastrophe to exactly the amount it felt comfortable paying were such a tragedy to occur, $500 million. Having left the market to develop nuclear energy for four years, the government became involved again, supposedly for the greater social good.

The amendments to the Atomic Energy Act in 1959 sought to stimulate nuclear power development "by a combination of three devices: a requirement of 'financial protection'; government indemnity; and an over-all limitation on liability" (Murphy, 1961, p. 16). Instead of allowing innovation in the market to develop safer applications of nuclear technology, as illustrated by appropriate insurance coverage, the government built on the philosophy undergirding the Price-Anderson Act and forced energy companies to purchase whatever insurance coverage was available and then provided the additional liability coverage up to a limit of $500 million. In addition, the government instituted the Power Reactor Demonstration Program to "subsidize the cost of fuel, conduct pertinent research and development work for free at the national labs, and help pay for some of the research and development conducted by private industry" (Morone and Woodhouse, 1989, p. 53).

By muscling into the development of nuclear technology, the government has altered the course of innovation in nuclear power production. When the Price-Anderson Act was passed, the size of the reactor was assumed to be quite large of necessity. This high risk (and price) associated with a nuclear reactor without government intervention would provide an incentive for individuals to develop and market new ways of containing and controlling nuclear power.

The Price-Anderson Act, argue Morone and Woodhouse (1989), preempted any such innovation and passed into law the ensured coverage of up to $500 million in the event of a nuclear catastrophe. Innovative applications of nuclear reactors may have resulted, for example, in the use of many small, dispersed reactors to generate power where the cost associated with an accident at any one reactor would be greatly reduced. Here, the federal government has subsidized high-risk large reactors and not allowed the market to develop methods of nuclear energy with acceptable levels of risk, as illustrated by adequate levels of privately provided insurance.

The Energy Policy Act of 2005 extended the Price-Anderson Act to 2025. In this current incarnation, power companies are required to purchase up to

$300 million per plant in *private* insurance. If there is a claim against any one of the country's 103 nuclear plants, every last one of them must contribute up to $95.8 million toward restitution for the damages caused by the offender. The power companies are thus required to insure each other, up to damages of approximately $9.5 billion. But power companies are insulated from any further damages they might cause. Previous experience tells us that if a catastrophe were to occur, with more money being needed, Congress would raise the money. But the important feature here is that the offending power company would not be liable for the additional damages. This is an invitation to what economists call "moral hazard." This refers to the phenomenon where an insured individual will undertake more risky behavior, simply because he is insured (e.g., you drive faster if you wear seatbelts). So the Price-Anderson Act encourages excessively risky behavior because the offending companies would not have to pay the costs of the damages they cause.

If the expected cost of a nuclear accident is less than $300 million, it pays the company to reduce its exposure, balancing the marginal benefit in insurance savings with the additional marginal cost of instituting the safety measures. The firm's marginal cost versus marginal benefit decision shifts toward risky behavior once expected losses rise to between $300 and $395.8 million. If the benefits of engaging in more risk (in terms of cost savings) are borne by the firm, but the costs of increased liability are split 103 ways, then there is a tendency toward augmented risk. Further, once a firm expects to inflict damages in excess of $395.8 million, it incurs no additional insurance savings, only the costs of instituting the safety measures. It will clearly, then, not be in the firm's interest to reduce its exposure to risk, once it contemplates expected losses in excess of $395.8 million.

If we experienced the type of severe nuclear accident that Greenpeace (Riccio, 2001) believes we could, damages would be in the range of $560 billion in 2000 dollars. This may be an overly pessimistic estimate, but let us use it for illustrative purposes.

If damages were $560 billion, the firm's liability would be $395.8 million. If damages were reduced to $400 million, the firm's liability remains unchanged. Why, then, would a rational company spend any money to reduce its liability to anything over $395.8 million. Such an investment would be pure loss for the firm.[2]

All insurance induces some moral hazard.[3] But because the Price-Anderson Act has completely limited the legal liability of a nuclear firm, that is, because it is subsidized insurance—it creates more moral hazard than would have arisen through the market. Instead of repealing the risk-inducing Price-Anderson Act, the government turned to a labyrinth of safety regulations.

We join Greenpeace USA (Riccio, 2001), the Cato Institute (Brownstein, 1984), and The Brookings Institution (Nivola, 2004) in opposing the Price-Anderson Act.

NRC REGULATION

Morone and Woodhouse (1989, p. 120) ask the question, "What about the possibility that the technology was effectively out of society's control because decisions were made by a nonrepresentative group of elites?" This is indeed what happened when the "group of elites" is understood to be government bureaucrats. They further describe the result of this involvement:

> There followed two decades of regulatory changes in reactor design and operation, but virtually all such modifications were variations within an unchanging basic approach. The momentum effectively precluded serious exploration of alternative technologies and significantly different approaches to safety. (1989, p. 125)

Their point concerns the multitude of possible types of nuclear reactors. As government regulation and licensing of private reactors increased, the specifics contained in those regulations had more and more to do with a single type of reactor, necessarily precluding any possibility of innovation down alternative avenues.

Currently, Section 50, Title 10 of the Code of Federal Regulations governs the licensing of nuclear reactors. In this 200-page exposition, specifics regarding a plant's "managerial and administrative controls to be used to assure safe operation" (§50.34.b.6.ii), "plans for preoperational testing and initial operations" (§50.34.b.6.iii), "plans for conduct of normal operations" (§50 .34.b.6.iv), "plans for coping with emergencies" (§50.34.b.6.v), and "proposed technical specifications prepared in accordance with the requirements of §50.36" (§50.34.b.6.vi) are all subject to the approval process. Basically, this code specifies the way a nuclear power plant must be run, down to the most minute technical specifications of the amount and placement of automatic isolation valves (§50 Appendix A Criterion 56.4). It is unfortunate that regulators and legislators cannot see into the future, or account for as-yet undreamed of alternative future technologies that might arise. Alas, we are all human. Current regulations are based on the amount and type of technology possessed today. If, for example, a new reactor was developed that did not require automatic safety valves, the reactor would not be approved for licensing without adding the extraneous valves.

Michael Porter and Claas van der Linde (1995) disagree with the claim that too much government direction stifles technological advancement, and instead argue that it can stimulate innovation. However, their idea that "regulation signals companies about likely resource inefficiencies and potential technological improvements" (1995, p. 98) is completely untenable. Buried in this statement is the assumption that the government somehow has knowledge of a "more efficient" resource allocation and benevolently informs entrepreneurs of their oversight via regulations that guide private industry down the socially beneficial road[4]. For example, Porter and van der Linde might imagine the NRC discovering that the benefits of a particular valve (in terms of increased energy production, decreased risk of accident, or whatever) are greater than the costs of that valve. Porter and van der Linde would then have the NRC mandate that nuclear plants use that valve. But how did the NRC discover this, while this was somehow kept hidden from the practitioners of nuclear energy production?[5] If we have learned anything from the fall of communism in 1989, it is that the owners of property have the most expertise in the application and preservation of their property, not a coalition of "experts" commissioned by the government. Further, if the benefits of the valve were indeed greater than its costs, wouldn't any profit-maximizing firm voluntarily choose to use the valve? If this mechanism really is cost-efficient, then firms will voluntarily choose to use it. If it is not, but the NRC mistakenly believes it to be, then the NRC has introduced greater inefficiency. In reality, an individual who sees an unexploited opportunity to profit from some resource allocation and chooses to act on this knowledge is an entrepreneur (Kirzner, 1973). The possible profits direct resource owners to apply their resources in the most efficient manner given the level of technology available.

Another claim by Porter and van der Linde (1995, p. 100) is that "regulation reduces the uncertainty that investments to address [the] environment will be valuable. Greater certainty encourages investment in any area." One problem with this claim is that it assumes that more "investment in any area" is better than less. But more investment in any one area *necessarily* implies *less* in some other area. It is unclear as to why there is a case for more investment in the regulated sector of the economy at the cost of less in the unregulated sector.

Another problem is that it is by no means true that regulation leads to less uncertainty that produces, in turn, more investment. Very much to the contrary, the possibility of regulation affecting the legality of investments *increases* uncertainty and thus *reduces* the amount of investments.[6] This is *very* much the case with regard to rent control, a price regulation on rental housing. This program has been so destructive of new building and upkeep of old dwellings, that even proponents of government regulation in general have acknowledged this effect. For example, stated Lindbeck (1972, p. 39): "In many cases, rent control appears to be the most efficient technique presently known to destroy a city except for bombing"; and according to Myrdal

(1965): "Rent control has in certain western countries constituted, maybe, the worst example of poor planning by governments lacking courage and vision."

Regulation may increase uncertainty and decrease investment for another reason. The limited length of time conferred by nuclear power plant licenses decreases the life of the investment and the value of it. A publication of the NRC illustrates this uncertainty: "the original operating license for a nuclear power plant has a term of 40 years. The license may be renewed for up to an additional 20 years if NRC requirements are met" (Strasma and Mitlyng, 2004, p. 1). A nuclear power plant requires a large amount of capital. Combined with the limited profit opportunity in the energy production industry, a long-term approach to returns must be adopted.[7] The limited life of 40 years and the uncertain extension of up to 20 years more gives an individual constructing a nuclear power plant, at best, 60 years to recoup and profit from the initial investment. The time frame may be too short, or too long, but the point is that the lifespan of a nuclear reactor is arbitrarily determined by the NRC and the operations of such a plant are just as subject to public opinion (if not more so) as they are to the profitable production and distribution of energy.

Consider the example of the Shoreham Nuclear Power Station in Long Island. The owners of Shoreham were granted a license to begin construction of the plant in 1973. After years of redesigns in order to comply with changing regulations, Shoreham was issued a license in 1985 to begin testing the plant at 5 percent capacity. That license was effectively revoked a few months, and $5.6 billion dollars, later due to intense public opposition. Not 1 kilowatt of commercial power was ever generated at the Shoreham plant (State of New York Office of the State Comptroller, Division of Management Audit). One detail of this story will serve as a counterexample to Porter and van der Linde:

> The Shoreham plant on New York's Long Island is a virtual twin to the Millstone 1 plant in Connecticut, both ordered in the mid-'60s. Millstone, completed for $101 million, has been generating electricity [to this day] . . . Shoreham, however, was singled out by anti-nuclear activists. (American Nuclear Society)

Certainly, if Millstone-1 was considered safe, why wasn't Shoreham? Contrary to Porter and van der Linde, regulations seem to have increased uncertainty, as the two plants are virtually identical. What was the difference that made the difference? Shoreham was too close to New York City. We grant that proximity to a major city should be a major consideration. However, this does not invalidate our point: regulations can change and completely wipe out the value of an investment. Investment uncertainty is improperly increased.

In what amounts to an astonishing admission, in 1999, the NRC decided to switch regulatory focus to a more risk-informed basis. The Nuclear Energy Institute (NEI)[8] seemed pleased with this, since the NRC would be "devoting less regulatory attention to issues with little or no safety significance" (NEI, 2004). In testimony before a Senate subcommittee, Gary Jones of the General Accounting Office reported that the "NRC believes that a risk-informed approach would reduce unnecessary regulatory burden and costs, without reducing safety." Balancing the costs and benefits, risks and rewards, is exactly what the market does best. Risk assessment is exactly what the insurance industry would insist upon. Profitable business for an insurance company is impossible without proper risk assessment. That is why they ask you if you smoke when you apply for health insurance; smokers are at a quantifiably greater risk of contracting lung cancer. Similarly, a properly priced nuclear insurance policy would require estimating the likelihood of various nuclear catastrophes.

The NRC, on the other hand, has not taken risk factors into account when it created its regulations. In the same senate subcommittee meeting, Gary Jones reported that

> Since the early 1980s, NRC has *considered applying risk to the regulatory process.* [Emphasis added.] To facilitate a discussion about the enforcement program . . . at NRC's request, NEI and the Union of Concerned Scientists reviewed 56 enforcement actions taken by the agency during fiscal year 1998. . . . From an overall perspective, the Union concluded that NRC's actions are neither consistent nor repeatable and that the enforcement actions did not always reflect the severity of the offense.

That enforcement did not always reflect the severity of the offense is very troubling, for this is exactly what we would hope the regulator would do.[9] Either enforcement was too lax, and therefore too many unsecured risks were being taken, or it was too stringent, in which case costs were unnecessarily too high. Neither case is an efficient outcome. The cost savings from more targeted regulation could be substantial, given that "In 1997, for example, 94 percent of violations had little or no safety significance" (NEI, 2004).

At least the NRC has, as of 1999, begun to address the issue of risk in its regulatory rule-making. Has it done a good job? Arguably, no. David Lochbaum (2000), writing for the Union of Concerned Scientists, conducted a study of the NRC's risk-assessment performance. He found that "the accident probability calculations are seriously flawed" (2000, p. v). Though he found many flaws, we will address here only two of them.

First, the probabilities of failure are assumed to be a constant function of plant age. For statistics aficionados, this means that failure "arrives" according to a Poisson distribution. But experience has shown that problems are less likely to occur for medium-aged plants; they typically arise when the plant is either brand new and untested, or when it is very old and rickety (Lochbaum, 2000). This oversight can easily be solved. Engineers have used computers to model U-shaped failure probabilities for decades, and non-Poisson distributions are commonly taught to actuarial studies undergraduates. It would appear that this is above the capabilities of the NRC.

> The most serious flaw is that the NRC has forgotten half the risk equation. If Green bets $100, and Brown bets $100 million, on a coin flip, the NRC would assess the risks to be equal in both cases; that is, the risk is 50-50. But clearly you are taking a bigger risk than I am. Risk, properly assessed, does not merely focus on probabilities. It focuses on expected costs. A 0.005% chance of catastrophic meltdown should be assessed a different risk factor if the plant is in Manhattan, than it is in Antarctica. This is common sense that the NRC has missed (Lochbaum, 2000). In contrast, insurance companies rely upon this consideration every time they write a new policy.

Section 50.11 of the Code of Federal Regulations deals with exceptions and exemptions from licensing requirements. It states that the "manufacture, productions, or acquisition by the Department of Defense" (§50.11.a), "performance of work for the Department at a United States government-owned or controlled site" (§50.11.1.i.A), and "research in, or development, manufacture, storage, testing or transportation of atomic weapons" (§50.11.1.i.B) are all exempt from the regulations to which private industry is subject. However, Benson (2004, p. 2) points out that U.S. Department of Energy nuclear weapons laboratory, production, and test facilities "have produced more than 99 percent of all the high-level radioactive waste in America." According to this claim, these regulations fail to limit the actions of the worst radioactive offender, namely the government itself. It is in this area that environmental activists misdirect their zealousness.

Doesn't this argue for more regulation? It certainly argues for some, but the salient point is, who is doing the regulating? The problem is that there is a conflict of interest when the government is both the creator and regulator of nuclear waste. This is why the Departments of Defense and Energy do not fall under the NRC's regulatory reach. The government, in effect, has exempted itself from its own regulations, and has polluted in the process (DiLorenzo, 1990A). If the government is to regulate itself, then we are right back where we started, with a conflict of interest. Under a system of private production, firms are largely regulated by the safety concerns of their insurance companies (Hoppe, 2001). Why not by the safety concerns of the government? Because regulation from this source tends to be a one-size-fits-all prescription, and changes in regulations are easily moved by political digressions and fickle public opinion. Private insurance companies competing with each other for the nuclear industry's business have an interest in responsive regulation that at the same time minimizes risk. If no insurance companies are willing to bear the risk of a nuclear catastrophe, then to the market's best judgment[10] the expected costs outweigh the benefits (i.e., the risks are simply too high). This is a market signal for either (a) withdrawal from the industry, or (b) greater research into nuclear safety technology.

Government has not allowed individual owners of resources to determine how best to allocate nuclear power in the framework of private property rights. In so doing, the government directed the development of nuclear technology by heavily regulating the industry. These regulations, due to their specifics, assume away any possibility of innovation. A given technology was understood at the time the regulations were drafted, and so forced nuclear advancement into a given paradigm based on this antique technology. Only where entrepreneurs are limited by their own imagination in the allocation of resources to relieve scarcity can innovation truly occur in the field of nuclear power and all others.

Nuclear Waste and Pollution

A major challenge to nuclear power is what to do with the spent fuel. Government has socialized disposal management services for the country, and atomic waste is no different. It is no surprise that the Hanford Nuclear Reservation, created as part of the Manhattan Project via government influence and involvement, "remains the nation's most contaminated site" (Dininny, 2004, p. 1).[11] The government-run atomic programs remain the worst offenders in terms of radioactive pollution. There exists constant conflict between a government deciding where to exercise eminent domain in siting nuclear waste repositories and communities chanting, "Not in my backyard!"

How can we best address the NIMBY problem? In the case of radioactive waste storage and disposal, the "solution of the problem is to be found in a lack of clearly defined or enforced property rights" (Cordato, 2004). Communities protest the possible siting of atomic waste repositories nearby because this process involves the government asserting eminent domain to acquire nearby property.[12] In addition, the state will not be held accountable for any violations of private property by subatomic particles (i.e., pollution) on the persons or property of nearby individuals because the "rules of product liability are suspended when the end product comes into the hands of an instrumentality of the United States" (Murphy, 1961, p. 14). The obvious solution is then to take this service away from the government and put it in the hands of responsible individuals acting in their own self-interest, and therefore the benefit of all. The NIMBY problem will no longer exist when the lands of all those who may be affected by such sites are purchased by private enterprise, and when neighbors no longer have the right to determine what is done on nearby properties, absent clear and present danger of invasions or other such trespass.

The NIMBY problem can also be solved by a restrictive covenant, which is in effect the market's version of zoning. When a property owner sells his land, he can stipulate in the contract that the buyer gives up his right to use his land in a particular way, or to sell to a specific party. In this case, it could be stipulated that the new owners do not resell to a nuclear power plant. Of course, this would entail large transactions costs if 10,000 neighbors were asked to sign such an agreement. There would be balking, fussing, holding out, and so on. However, if there were an initial single owner of a large plot of land, big enough to encompass the said 10,000 housing plots, this entrepreneur could earn greater profits by tailoring a restrictive covenant to the tastes of his customers. For example, he could sell them their holding subject to the condition that they never open a nuclear power plant on their land, nor vend to anyone else without imposing this condition on them. The land with this "restriction" would actually be worth more than in its absence.[13]

Granted, a restrictive covenant of these 10,000 people cannot prohibit adjacent landowners from installing this dreaded technology on their own property. The beauty of the market in this regard, however, is that transactions costs are a lot lower when groups composed of 10,000 homeowners each, that is, homeowners' associations of this sort, negotiate with each on matters of this sort.

If we dispose with the regulations governing nuclear waste disposal, wouldn't the nuclear companies dump their radioactive waste anywhere, thereby irreparably polluting the environment? Again, the answer depends on the degree to which the free market is able to function. If land is held in common, as is much of the world's oceans, then a common economic problem called the tragedy of the commons arises. The tragedy is that if the benefits of polluting are borne by the individual polluter, but the costs are split evenly among the population (since the polluted land is common property) then there is a tendency to over-pollute. Where private property is concerned, people are far less likely to pollute their own land.[14] If a nuclear power company wanted to store its radioactive material on a site it owns, it is free to do so. It can do what it wants with its land. But it does not have the right to pollute other people's land. Thus, if some of the radiation seeps out onto the holdings of other people, a violation of property rights has occurred. The private owner can then take matters into his own hands, and sue the polluter; he would certainly be justified in obtaining prior restraint or a legal injunction. Or, the company might consider paying the private owner for the right to pollute his land to a mutually agreed-upon level.

Let us make a more philosophical rebuttal to the zoning challenge. Under the libertarian legal code (Rothbard, 1982; Kinsella, 1996), the only time force is justified is to respond to, or defend against, or punish, a prior use of force or threat. A zoning law forbidding the construction of a nuclear facility without prior approval would thus be illegitimate. The only situation under which such an edifice could be properly restrained before completion would be if it constituted a threat: a clear and present danger. But, given the exemplary record of this industry, even under government control, no rational court could make any such finding.

A referee of this Journal asks "Why can't nuclear plants constitute threats?" The essence of a threat is, well, to *threaten* someone if they do not do as you wish. For example, in robbery, extortion, the threat might be couched as follows: "If you do not give me money, you will be shot or your children kidnapped." Setting up a nuclear generator, obviously, has nothing to do with anything even remotely connected to this. But there is a peripheral or extended meaning of threat that does encompass acting in a negligent manner, one likely to cause an accident. For example, Jones shoots high-powered arrows at a target set up on a tree on his own property. If he misses, the arrow will land on Smith's property, and, quite possibly, do great harm there. At first glance, nuclear power might well fall under this heading. After all, if one of these installations blows up, it is an extreme understatement to say that massive destruction will ensue. So, what is the record of this industry, even under (we contend inefficient) government control? Pretty good, actually. Perhaps that famous bumper sticker says it all: "More people died at Chappaquiddick than at Three Mile Island."[15] Another consideration: if atomic generators are declared a "threat" and banned by law on that account, then so must the entire airline industry, which has killed far more non-passengers.[16] We regard this as a reductio ad absurdum, as no one, no one, advocates grounding all planes for this or any other such reason.

There need not be any fear that nuclear power stations would be located under free enterprise, cheek by jowl with high population concentrations. For one thing, insurance would be far more expensive. For another, so would real estate purchases. The zoning which is part and parcel of the free enterprise system (Siegan, 1970, 1972) would thus tend to ensure that such facilities would be located in virtually empty locales. The market zones itself. Consider what it would cost to purchase on the free market enough space for a private garbage dump in Manhattan; or an automobile manufacturing plant there. These things do not occur in a free market because of the opportunity costs associated. To purchase land in Manhattan, or any such residential or business area, is far more costly than land in uninhabited, remote places.[17] By and large, factories are founded in the periphery. Only after they have been established do private individuals decide whether it is in their self-interest to build homes in the neighborhood. If people wish to buy land next to Yucca Mountain, we do not oppose this.

This is not to deny, of course, the possibility that neighboring land-holders in these out-of-the-way places might possibly suffer a loss in property values unrelated to intrusion by subatomic particles. Who, after all, wants to be located next to a nuclear power plant? However, in the libertarian legal code, one does not own the *value* of one's property, only its physical integrity (Hoppe and Block, 2002). Thus, protests on NIMBY grounds, for this reason, would be ruled out of court.

The possibilities of nuclear energy have been held in check indefinitely by government regulation. There is no reason to believe that nuclear power plants must be as large in scale as coal plants. Yet, this view was adopted when drafting the current regulation regarding nuclear reactors because of the technology of energy production and distribution at the time. A drawback to large nuclear plants is the tremendous liability due to the possibility of an atomic accident. Although it is difficult if not impossible to anticipate the future workings of a fully free market, in a truly privatized energy industry, a network of small reactors may minimize risk.[18] As to the radiation of the actual power plant, "[nuclear power] produces no carbon dioxide and radioactive emissions are actually *lower* than the radioactivity caused by coal-fueled power plants" (Lomborg, 2004, 129). The greatest problem the public may have with a nuclear power plant is the possibility of a major accident; however, "in the last forty years . . . not one single fatality has occurred as a result of the operation of a civilian nuclear power plant in the United States" (Electricity: Benefits / Effects, 2).[19] And this is with outmoded and inefficient Atomic Energy Commission in charge of the industry's safety. Under a pure profit and loss-free enterprise system, experience suggests safety will be improved.

The Fate of the Nuclear Industry

Should there even exist a nuclear power industry? What about Chernobyl? Chernobyl was the world's worst nuclear disaster. Estimates of the death toll vary widely. Industry estimates that thirty-one people were killed (Gonzalez, 1996). Other estimates are "4,000 deaths among those who took part in the hasty and poorly organized cleanup; 70,000 people disabled by the radiation" (Greenpeace, Probability of a Nuclear Accident, p. 3).

In a study, "Chernobyl—Ten years after" (Gonzalez, 1996) a group of more than 800 experts from 71 countries came to a consensus view about the effects of the disaster. Their findings pointed to significant thyroid problems among children in the region. The study concluded:

> The radiation levels that can still be detected in most affected areas are sufficiently low as to permit normal economic and social activity to be resumed. The health effects have not turned out to be as catastrophic as some feared and others reported. But a number of radiation effects did occur and more are expected to occur and should be dealt with. Moreover, the socioeconomic impacts are very serious.[20]

The problem becomes, how might one weight the costs and benefits of environmental damage, versus health damage, and so on? Are these costs high, or low? Without some common denominator, such questions are impossible to answer.

Yuri Koryakin, a Soviet nuclear industry economist, estimated that "by 2000 . . . the Chernobyl accident would cost the country between 170 and 215 billion rubles from contaminated farm land, lost electricity production and other economic consequences." This is far in excess of the net economic contribution of the entire Soviet nuclear industry, estimated at between 10 and 50 billion rubles. In terms of costs and benefits, "the Soviet Union may have been better off if it had never begun building nuclear reactors in the first place" (Hudson, 1990, p. A8).

Clearly, such cost benefit analysis is necessary, but the real cost/benefit analysis can truly take place only in a free market. Such argumentation was the centerpiece of Ludwig von Mises's (1969) critique of socialism. How can anyone know how many resources to put into an industry, if there are no prices? How can one estimate the cost from depleted fish stocks, and add them to the costs of health damage? You can only do that if these costs are denominated in the same units. Only if these costs take on dollar terms, and can they be quantified, aggregated, and thus compared. Market prices indicate relative scarcities and surpluses. They tell us if a resource would be better used elsewhere. Should steel be used to make a car factory or a nuclear power plant? We can only decide such questions if we know the price of steel, the price of constructing the two plants, and the expected revenues from auto and energy production. It is only to the extent that private property rights permeate an economy that these questions can be answered.

Insurance companies estimate the potential risks in pricing their policies. These prices are factored into the cost calculations of a nuclear firm. If the sum of all these costs is greater than the expected benefits of energy production, then it will not pay any profit-seeker to build nuclear power plants. The problem for regulators is that they cannot beat the performance of the insurance companies. Experience suggests that their risk-assessment capabilities are far inferior to those of profit-seeking insurance companies. But even on the best-case scenario that the government can just as effectively quantify the risk as can the insurance companies, this still does not mean that government should be doing the regulating. Why not? For one thing, only if the insurance policy is priced by the market, can anyone know the proper level of insurance (i.e., how much insurance to buy). Just because government can equally well quantify risk does not mean that their pricing policy will not be high jacked by political and special interest considerations. For another, even if impossible the government would replicate the market, what is the case for it doing so? Surely, on the grounds of specialization and division of labor, the state should cease and desist under these conditions.

How much nuclear power should there be? Lorna Salzman (1979, p. 112), arguing against nuclear power, unknowingly stumbled onto something when she claims that "*no one knows all the questions that must be asked and answered* in order to find solutions." This is precisely what economists mean when they say that "competition is a discovery procedure" (Kirzner, 1985). We cannot know ex ante what the best allocation of resources will be. We don't know everyone's costs—in fact, costs are inherently subjective and as such, they are unknowable by anyone else. Similarly we must necessarily be in ignorance of the benefits of a particular economic outcome. However, the market provides a forum where everyone's costs and benefits are addressed in the form of supply and demand. For this reason, government cannot be relied on to direct private industry to the "social good" or to minimize "externalities" because it cannot know what that good is or how these externalities, if they exist in the first place, can best be internalized. It cannot learn the appropriate technology required to ideally serve the public. Innovation, almost by definition, cannot be planned.[21,22] Anti-market environmentalists such as Salzman find themselves railing against the government whether they like it or not when they protest the storage and disposal of spent nuclear fuel, because only the government has been legally allowed to perform this service.

Thus no one can answer the burning question: Should there be any nuclear power? Only the market can answer that.

CONCLUSION

Let us summarize. The nuclear industry was too risky for private firms to enter without subsidization. That is, the risk was too great for any company, and no insurance companies would shoulder this burden either. Government wanted nuclear energy so it limited, via the Price-Anderson Act, the liability of the nuclear industry. The Act enticed firms to enter the industry that would not have done so otherwise, causing an artificially low price for its energy. The Price-Anderson Act causes a moral hazard situation, which encourages further risky behavior, increasing the likelihood of a catastrophe, the costs of which the offending firm would not have to pay. How did we address this problem? Instead of dropping the Price-Anderson Act, the government opted instead for a regulatory system that was, from day one, fundamentally flawed, and continues to have major problems. To the extent that nuclear operators spend too many resources dealing with non-safety-related paperwork, the costs of nuclear power increase. So the industry faces price distortions in both directions. Finally, why did the government renew the Price-Anderson Act? It was because of the very risky behavior of nuclear operators which the Price-Anderson Act encouraged. This is clearly circular logic; the government's inefficiency used as a rationale for more governmental inefficiency. This is a road to serfdom (Hayek, 1944).

The present chapter is neither pro nor antinuclear power in terms of its public policy proposals. Rather, the view espoused in this quarter is that there should be an "even playing field"[23] between *all* sources of energy: the mainstream ones, such as coal, oil, and natural gas, the gaia-ist ones such as water (hydro-electric), wind, and solar, and the "conservative" one under discussion, nuclear power. But the competition between them can only be "fair" if each one pays its own way; if none of them are allowed to violate the property rights of third parties (Rothbard, 1982). The Price-Anderson Act should be repealed, forthwith, along with all other legislation that gives this industry an unfair advantage. As well, subsidies to it should also be consigned to the dust-bin of economic history, so that nuclear power can succeed or fail entirely on its own merits. The same ought to be done with regard to all competing sources of energy. Then and only then will the total benefits and costs (including health and environmental costs) be allowed to determine the contours of the energy industry.

Right now, paper clips compete with staples, glue, and rubber bands on a more or less even playing field. The government gives no special favor and imposes no particular fear on any of these industries devoted to keeping paper in order. Thus, there is a certain rationality involved with them. We pretty much know that the allocation of resources between paper clips, staples, glue, and rubber bands is a reasonable one. Were it not, for any reason, market forces would soon enough bring matters back into conformity with consumer tastes, given relative costs. Would that a similar situation prevail in the energy field, and the political economic problem with regard to it would pretty much vanish.

NOTES

1. The government, surprisingly, failed to recognize that the monopoly it granted to energy companies at the time was the source of sluggish development. Though established in the field of energy distribution, there is no reason the production of

energy from nuclear sources would follow the same industrial organization of other forms of energy production. Development by companies outside the current circle of privileged energy producers did not exist because any other firm other than current energy utilities could not benefit from the use of the new technology.

2. This is exactly the same type of reasoning that some people have used against mandatory death penalties: a criminal on the run, who knows he will be executed if he gets caught, has every incentive to kill anyone who stands in his way.

3. Under the pressure of marketplace competition in this industry, however, this is kept to a minimum.

4. Hayek (1989) called this the "fatal conceit."

5. We do not deny that sometimes, rarely, government does discover something before private enterprise. Radar is a case in point. However, there is a weeding out process in the latter case that simply does not function in the former: profit and loss. Thus, over the long haul, we can expect, and empirical evidence bears this out, that entrepreneurs are far more likely than bureaucrats to make commercially viable breakthroughs.

6. For the more general argument against rent control on the ground that it *reduces* investment in residential rental units, see, Baird (1980), Block and Olsen (1981), Block, Horton and Shorter (1998), Block (2002), Grampp (1950), Hayek (1981), Johnson (1982), Lindbeck (1972), Myrdal (1965), Salins (1980), Tucker (1990).

7. This is due to the application of anti-trust laws on energy utility companies. Companies in the energy industry must walk a thin line between predatory pricing, collusion, and price gouging.

8. This organization is to the nuclear industry what the NRA is to guns.

9. Were we, that is, to rely on government regulations in the first place, which we do not.

10. To use a crass anthropomorphism

11. For the case that government is the worst polluter of them all, see DiLorenzo (1990A).

12. Though we consider all exercises of eminent domain to be illegitimate, this critique would take us too far from the topic at hand. For a critique of eminent domain, see Epstein and Block (2004), Block and Block (1996), Block (1979, 1998).

13. Suppose that in 100 years nuclear power plants become no more dangerous than are Frisbees nowadays, and that people, then, want to have their own backyard installations of this technology. Will the dead hand of the past, these restrictive covenants, prohibit them from doing so? Not at all. The members of the covenant can agree beforehand, that is, now, that a certain proportion of them can always overturn any specific element of the contract.

14. Indeed, this is a contradiction in terms. One, logically, *cannot* pollute his own property. He can of course use it as a garbage dump. But it is not *pollution* until it lands on someone else's property.

15. See also fn. 33 on this point.

16. We ignore that meltdown that occurred in Chernobyl; Russian technology is vastly inferior to our own (see text at fn. 34 for more on Chernobyl). We do not

count the tragic deaths of 9/11; that was purposeful, not accidental. We abstract from airline passengers who died in crashes; they voluntarily took upon themselves the risk of travel in this mode. We focus solely on "civilians," people on the ground who perished in air crashes.

17. This is the same argument, in the other direction, for why the free market would provide for public goods. The positive externalities associated with waterfront public sidewalks is just the converse of the negative externalities of public nuclear plants. The detailed argument can be found in Block and Block (2005).

18. In Rifkin (2002) the possibility of a worldwide energy web is discussed. While he applies this idea only to the production of electricity by hydrogen fuel cells, there is no reason why it cannot be applied to other forms of energy production. A network of nuclear reactors the size of submarine engines may produce the same amount of energy as one large facility, with much less liability in the case of a cataclysmic event. This is one way the private insurance problem may have been overcome had government allowed the private enterprise to determine for itself how to solve the problem of making new technology viable for the market. For more on the benefits of the insurance industry to free enterprise, see Hoppe (2001).

19. A referee has pointed out that this statement is controversial. The consensus is that, even at Three Mile Island, no one died. However, in his 1981 book *Secret Fallout*, professor of radiation physics Dr. Ernest Sternglass, who estimated that more than 430 infants died as a result of low-level radiation poisoning. Allen Lutins reports six workers have died in non-research power plants, five of which were from non-nuclear causes such as steam scalds and sabotage. Suffice it to say, that even if these two gentlemen are correct, the risk of injury is extremely low in the nuclear industry. The American Nuclear Society reported that accidents which resulted in lost work time plus fatalities was 0.13 per 100,000 worker-hours for nuclear versus 1.55 for U.S. private industry in general.

20. We thank an anonymous referee for pointing us to this very important research.

21. A government that initiates a set of regulations regarding an industry bases these regulations on a given technology known at the time. The possibility of innovation in this technology cannot be taken into account by the regulation, and the regulation necessarily precludes such innovation from existence in that future technologies must be subject to the regulation based on the historic or outdated technology.

22. It cannot be denied that sometimes, in a rare while, innovations stem from governmental activity. Radar came about from World War II efforts; some people credit the U.S. military with early contributions to the internet; certainly, the moon shot was a statist operation to its core. However, we may always ask the question of whether these breakthroughs came about because of or in spite of governmental activity. That is, had the state not diverted to its own coffers gigantic amounts of treasure from the private sector, might these innovations come about sooner, or better, or more efficiently?

23. To adopt that silly phrase from the debates over free trade.

Chapter 4

Energy Petition Memo to the NTU

From time to time, I am asked to sign petitions, mostly of a free market orientation (those that oppose economic freedom are quickly consigned to the delete file). Whenever I can, I do sign them; I am convinced that they do no harm, may do some good, and take very little time to read and add my name to them in any case. So, the opportunity costs are small.

But, every once in a while I decline, because while I need not enthusiastically support each and every word in a given petition, I am uncomfortable signing my name to any that violate what I take to be libertarian or Austrian economic principles.

The most recent one to cross my desk, well, be sent to me via email, was from the National Taxpayers Union. This is almost always a group that can be reliably counted upon to support capitalism. However, this petition unfortunately had several problems with it. I suggested a few changes that would allow me to hop aboard, but, usually, it is far too late in the process to make any changes after it is widely circulated, and many people have already signed on.

In what follows, please see the following: I. A letter to me from the NTU; II. The petition they had asked me to sign in this particular instance; and III, my response. See if you can pick out the problems I might have had with this missive before looking at my critique of it. (No peeking, now.)

THE LETTER

From: NTU
Sent: Mon 8/20/2007 4:21 p.m.
To: NTU
Subject: Sign-on to NTU Economist Letter on Opposing Energy Legislation
Sign on to NTU letter opposing anti-market energy legislation

Dear Economist:

You recently received an email from the National Taxpayers Union (NTU) asking you to sign on to the following statement on energy policy (see pasted copy below). The statement, which will be signed by economists throughout the nation, argues that the energy legislation currently before Congress contains yet another package of harmful taxes, regulations, and subsidies that will further distort energy markets in this country. Over 130 economists have already signed on in support. The statement will be released to the media and on our website later this month.

We have yet to receive a response from you and hope that you will consider signing on. If you would like to sign, all you need to do is send a brief reply email to NTU's Keith Capp indicating your title, name, organization, and your desire to sign on. (Note: Affiliations will be listed for identification purposes only.)

Thank you very much for your consideration of this important request.

Sincerely,

National Taxpayers Union

THE PETITION

An Open Letter to the U.S. Congress:

We, the undersigned economists,* write to strongly advise against the inclusion of damaging anti-market provisions in the energy legislation now moving through Congress. History has shown that attempts by the federal government to tax, regulate, and subsidize our way to more plentiful and secure energy have failed miserably. This Congress ought not to repeat those expensive mistakes when considering energy policy.

If Congress seeks a reduction in America's dependence on foreign oil, the tax increases found in the House's energy bill won't achieve the goal. Vengeful tax hikes on the so-called "Big Oil" serve no economic purpose. In 1980, Congress instituted a windfall profits tax to punish the industry. The result, according to a Congressional Research Service study, was a drop in domestic oil production of 3 to 6 percent and an increase in oil imports of 8 to 16 percent.

But instead of simplifying and equalizing tax treatment across all industries, current proposals attempt to bestow favors upon certain sectors at the expense of others. The energy bills propose dozens of new tax credits for the so-called "alternative energy" programs, many of which are expensive and not commercially viable.

The House and Senate bills also propose harmful mandates on fuel usage and energy production. Forced consumption of 36 billion gallons of heavily subsidized ethanol by the year 2022 will significantly raise fuel prices, taxes, and food costs for millions of consumers. Furthermore, requiring that electricity producers derive 15 percent of their power from alternative sources will artificially increase utility bills for American families.

Finally, the legislation purports to let government lead the way to energy independence by creating a slush fund from which alternative energy programs will be funded. Unfortunately, Congress cannot consistently identify and nurture emerging technologies because it distributes funding based on political concerns rather than sound science or economics. Just one example is the $2-billion Carter-era, feel-good boondoggle called the "Synfuels" program, which sought (and utterly failed) to produce alternatives to petroleum. The lesson from this and other experiences is that political forces cannot replicate the results of market forces.

If Congress wants to pursue a sound energy policy, it ought to reduce government interference in the markets that are capable of delivering innovative energy solutions to consumers. By easing regulatory burdens, ending distortions that divert productive capacity, and allowing the price mechanism to do its job, Congress would be providing American consumers and taxpayers with far more plentiful energy at a lower cost than any package of taxes, regulations, and subsidies could ever hope to do.

Sincerely,

The Undersigned (*Affiliations listed for identification purposes only)

MY RESPONSE

Dear NTU:

Sorry, I can't sign your otherwise very good petition, due to these three problems:

1. You say: "If Congress seeks a reduction in America's dependence on foreign oil,"

But this implies it is a reasonable goal for Congress to aim at, and that this bill just goes about attaining it in the wrong way. I don't see this as a reasonable goal. Why should Congress or anyone else for that matter have a say in the freely made decisions of the American people in determining the proportion of foreign and domestic oil they wish to consume?

However, the argument is made in some quarters that Americans purchasing foreign oil helps support dictatorships abroad. Why is this wrong?

It is wrong on several accounts.

First of all and perhaps most important, the United States is the world's biggest "dictator," not domestically but certainly in a foreign context. That is, the United States is a dictator to the peoples of numerous other nations. This being the case, following this principle, we should not be purchasing domestic oil. But if we cannot buy oil from either foreign or domestic sources, we are in dire straits. Only the most fervent anti-human environmentalists would welcome such a scenario (even though they do not act consistently with it in their own lives).

Second, it implies that the best way to promote liberty in those countries that are ruled by dictators is by boycotting them. This has not worked out too well in Cuba, has it? No, a more reasonable and efficacious way to deal with people in these nations is through commercial interaction. When we get to know them, and they, us, when bonds are forged between countries, dictatorship lessens. Free trade, full free trade, is not only economically beneficial, it helps with these sorts of challenges as well.

Third, a boycott hurts not so much the dictators of these countries, but the average citizen. Remember, the people who live in these foreign dictatorships are for the most part victims, not perpetrators of evil. How does an attempt to starve them help them? The hope is that this will encourage them to overthrow their overlords, but this has rarely ever taken place in history. Revolutions are typically waged by the middle or upper classes, not by starving people. If there is to be hope of change, leaders of places like Cuba most see the United States up close and personally. They must witness how much better their lives would be under our economic system than theirs. But this can only occur if barriers to trade, investment, and travel are reduced, not increased.

2. In the petition, the following words appear: "Unfortunately, Congress cannot consistently identify and nurture emerging technologies because it distributes funding based on political concerns rather than sound science or economics."

First, it is *impossible* for Congress to identify any such thing. Have we learned nothing from the fall of the USSR? I suggest the authors of this petition read Mises on the socialist calculation debate on this matter.

Second, suppose that we ignore this point. It would *still* be incompatible with libertarianism for Congress to engage in central planning even if it could.

Third, I think it is "fortunate" not "unfortunate" that Congress cannot do this.

3. The NTU petition also contains the following: "By easing regulatory burdens, ending distortions that divert productive capacity, and allowing the price mechanism to do its job, Congress would be providing American consumers and taxpayers with far more plentiful energy at a lower cost than any package of taxes, regulations, and subsidies could ever hope to do."

No, under these conditions, it wouldn't be *congress* that would be providing all these great things; rather, it would be the free enterprise system that would be doing so.

Rectify these three problems, and I'll gladly sign on.

Best regards,

Walter Block, Ph. D.

Chapter 5

Oil and Gas Stations

GAS STATION FOOD SALES AND BICYCLE RICKSHAWS

When foresighted entrepreneurs first offered the services of supermarkets and department stores in the modern era, they were greeted with terror and loathing by the smaller merchants already in operation.

Why this hysterical reaction to what was only a new form of commercial organization? The small shopkeepers feared, and rightly so, that the new business organization would attract their customers away.

As a result, they contrived to prohibit these retailing innovations and succeeded for many years. They bamboozled the public, convincing people that the "family grocery," like the "family farm," was the backbone of civilization. Without it, morals would collapse, and a new dark age would be ushered in. With the help of sympathetic bureaucrats, zoners, town planners, and politicians, they were able to impede and obstruct the giant retailing enterprises.

Happily, however, their efforts were eventually stymied. As a result, supermarkets were able to compete for customers, just like any other type of business. This was a great boon to the general populace, one that we now take for granted. Consumers benefited from the convenience of one-stop shopping and from the lower prices made possible by attendant economies of large scale. And there are still mom-and-pop grocery stores in existence, successfully competing with their gigantic counterparts.

(The major exception to this general rule is Japan, where small merchants associations have been more successful in handcuffing the shopping center idea, much to the impoverishment of the average person living in that country.)

Unfortunately, this attack on the consumer interest has not yet been put fully behind us in Canada. There are still merchants and their associations who actively attempt to restrict the commercial activities of competitors. They are ostensibly devoted to free enterprise, but in reality are its most bitter enemies. There are still civil servants who see their main goal in life as one of interfering, not supporting, the economic liberties of consumers and citizens.

Consider in this regard the laws of several municipalities, such as North Vancouver, B.C., which refuse to allow filling stations to sell hot food, or grocery products such as bread, milk, and laundry soap.

Mr. Percy Melville, that city's property use inspector, discovered three garages which contravened the municipal zoning bylaw by selling these prohibited grocery items. Letters were sent to the offenders, demanding that they either comply with the regulations, or file an application for permission to rezone their properties. According to city planner Pauline Jang, Shell agreed to comply with the existing bylaw regulations by phasing out prepared food, while Mohawk and Chevron, the other two violators, have decided to apply for a site-specific CD zoning variance, which would allow them to carry on as at present.

But wait a minute. We're not talking about purveyors of pornography or heroin here. These service stations are guilty of selling nothing more weird and exotic than bread, milk, and prepared food, for goodness sake! Why enmesh these property owners in reams of bureaucratic red tape?

And it is all so unfair. In North Vancouver District, the municipality right next door to the City of North Vancouver, competing filling station owners are allowed to sell any grocery items they wish. Whose interest is served by the prohibition and who is injured? Certainly the consumer is harmed, as are the service station owners who must compete with one hand tied behind their back. The winners are the merchants who have been able, through political pull, to save their market from the pressure of competitors.

However, the aldermen (alderpersons? alderpeople? aldercritters?) of the City of North Vancouver are supposed to legislate on behalf of all their constituents, not just for the squeaky wheel sector of the business community which abhors competition. It is not clear under these circumstances why the local government finds it appropriate to bring about a modern version of the medieval guild system.

Commercial self-interest (a "right-wing" sin against the market) is not the only motivation behind this anti-competitive philosophy. Left-wing ideological considerations play a part as well. Consider Vancouver alderman Harry Rankin's virulent opposition to bicycle rickshaws on the profoundly racist ground that they were "demeaning," and anti-working people. Due to popular sentiment, this mode of transportation was legalized in Vancouver, but only during the Expo era. At present, this non-traditional industry, non-traditional in Western culture, that is, is severely restricted.

And what a shame. Now, with the good weather soon to be upon us, there will be thousands of potential customers for the bicycle rickshaw. And there will be hundreds of young athletes running around on the streets, highways, and byways of our community, in any case, getting in shape for the next Sun Run, marathon, or triathlon. Many of them will likely be unemployed or underemployed students. Wouldn't it be great to legalize this activity, on a for-profit basis? Then, the free citizenry of B.C. could engage in some capitalist acts between consenting adults, for mutual advantage.

BEAUFORT OIL

The Inuvailuit Environmental Impact Review Board has just rejected a proposal by Gulf Canada Resources to begin a three-year exploration/drilling program. They did so because the company offered no evidence that it could counter an oil well blowout of Valdez proportions any better than was done at Prince William Sound, Alaska, in that case.

From one point of view, this is all well and good. If no one drills for oil in the Beaufort then, by definition, there can be no despoliation of the environment. On the other hand, if the next 100 or so applications for oil drilling all meet with this "Valdez phobia," there will soon be a time when we really will run out of oil resources. Appropriate public policy implies a careful and delicate balancing of these two downside risks. Unfortunately, present institutional arrangements are not at all designed to achieve such balance.

Look at it from the point of view of the Inuvailuit Environmental Impact Review Board. This agency, created as a component of the land claims settlement deal between Ottawa and the Western Arctic Inuit, has every incentive to take zero risks with the environment. If the drilling succeeds and no oil spills, the group responsible for approving an environmental impact statement earns no money; thereby, if there is a blowout, this group will have failed in its mission.

The amazing thing under these conditions is not that the review board failed to approve of the Gulf program; the amazing thing is that any review board, anywhere, might one day approve a similar program.

The answer to the allocation of risk problem is to ensure that the group responsible for such decisions gains, financially, if oil is safely developed and loses revenues if it is not. Traditionally, it was the institution of private property rights which was able to internalize the externalities in this way.

The only hope for safe development of the Beaufort Sea and other such resources may be a return to the doctrine of private property rights.

Section IV

PROPERTY RIGHTS

Chapter 6

Homesteading[1]

Richard Stroup (1988)[2] attempts to show that the Homestead Act of 1862 was an economic disaster, in that it led to needless misery and suffering on the part of people who settled the Western lands under that program. In his analysis, homesteading encouraged the too rapid pushing back of the frontier, dissipating all or virtually all of the economic value of these lands. From these comments he draws the conclusion that government auctions of virgin lands are preferable to homesteading them, a lesson with far more than only historical interest.

I shall begin by contending that the evidence offered for his finding is insufficient; I shall then put forth the view that the theoretical underpinnings of his argument are faulty, considering and rejecting several implications of the Stroup hypothesis; I conclude that it cannot rationally be applied to problems of a similar sort facing society today.

But before we begin, a preliminary question must be squarely faced. There is some question as to whether Stroup is analyzing and criticizing the (Lockean)[3] principle of homesteading, the homesteading era, or only the Homesteading Act of 1862. In my view, his remarks apply to all three, not merely to the latter. True, he begins by focusing on that piece of legislation, but his analysis is then applied to homesteading in general. Accordingly, I shall interpret his remarks as a critique of homesteading in general, as well as one of the law of 1862.

ECONOMIC HISTORY

What documentation does Stroup offer for his thesis of premature settlement? He only shows that the lives of these settlers were one of hardship, sacrifice, and of times failure to maintain their homesteads. But this undisputed fact is hardly a necessary or sufficient condition for resource misallocation of the sort he asserts took place. After all, this was the decade of the 1860s; suffering and adversity were by no means a monopoly of these frontier farmers. Numerous other people in the mid-nineteenth century led lives of extreme destitution and poverty.[4]

Further, what limited bankruptcy statistics that are available fail to indicate any massive increase attendant upon the 1862 law. For example, consider the 20-year period from 1857 to 1876:

It is very difficult to reconcile the Stroup hypothesis with this data. In the year before the passage of the homesteading legislation, American business failures reached a high point of 6,993. But with the advent of this law, bankruptcies began a sharp decline. It was not until 1875 that the number of insolvencies exceeded the level attained in 1861. And one would have to search far ahead[5] in history, all the way to the depression year of 1893, to exceed the level of losses that was registered in 1861.[6]

Yet another explanation of the impoverishment of these farmers—such as it was—has to do not with homesteading per se, but rather with the artificial size barriers placed upon this process by the government. In this regard states Murray Rothbard (1986):

Table 6.1

Year	*Number of Business Failures*	*Losses in Millions of Dollars*
1857	4,932	292
1858	4,225	96
1859	3,913	64
1860	3,676	80
1861	6,993	207
1862	1,652	23
1863	495	8
1864	520	9
1865	530	18
1866	1,505	54
1867	2,780	97
1868	2,608	64
1869	2,799	75
1870	3,546	88
1871	2,915	85
1872	4,069	121
1873	5,183	229
1874	5,830	155
1875	7,740	201
1876	9,092	191

Source: Historical Statistics of the U.S., U.S. Department of Commerce (1975)

> Destruction of the grass cover in the West in the late 19th century was caused by the federal government's failure to recognize homesteading of land in large-enough technological units to be feasible. The 160-acre legal maximum for private homesteading imposed during the Civil War made sense for the wet agriculture of the East; but it made no sense in the dry area of the West, where no farm of less than one or two thousand acres was feasible.[7]

And there is a further reason to criticize the Homestead Act of 1862, but not the Lockean notion of establishing property rights in land by mixing one's labor with virgin territory: the enactment placed onerous roadblocks in the path of the would-be homesteader which were not called for by the theory itself. Anderson and Hill (1983, p. 445) describe the requirements of land claims clubs and associations which functioned in the Western states in the nineteenth century, and were run roughly along the lines called for by Locke: "It should be emphasized that these expenditures were specified in terms of labor, allowing the farmer a great deal of latitude with his investment. . . . This is in sharp contrast to the Homestead Acts which specified the investments in terms of cabin size, irrigation ditches, and planted trees, most of which were unnecessary."

How could we put our alternative hypothesis in terms of Stroup's figure 1? One possibility would be to allow his function to begin at the origin, be coincident with the x-axis until point t*, and then continue to the right as he has depicted it. In his terminology, this would be equivalent to denying the applicability of the "farm frontier" to this particular historical epoch.

I must conclude this first section with the claim that Stroup has failed to demonstrate that the Homesteading Act of 1862, let alone homesteading per se, was the cause of excessive suffering on the part of the settlers. It may possibly be true, but it has not clearly been shown to be true. As a pure matter of economic history, the best that can be said for this hypothesis is that it remains unverified.

ECONOMIC THEORY

Even if Stroup has not shown his theory to have had application to the post–civil[8] war period, it is still a very interesting one. Could his critique of homesteading in general still be viable, and thus potentially of relevance at some other time? In this section, we shall suppose that this is so; in other words, we shall for argument's sake accept the "farm frontier" model, and thus figure 1 exactly as it appears. We now ask if there are any additional and special conditions still needed for his policy analysis to be correct.

Even if we fully accept his critique of homesteading, it by no means follows that a government auction[9] is a preferable public policy. One need to do more than demonstrate "market failure" in order to justify government intervention; one must also defend the position that public-sector involvement will not bring in its wake even worse implications for economic welfare. That is, one must show that the "cure" will not be worse than the "disease." But Stroup fails to do this, implicitly assuming that his demonstration of market imperfection is sufficient to warrant the statist auction system.

One way of looking at this issue is to see that what is at stake here is the issue of privatization. Homesteading is the market's means of converting virgin land into private ownership status; the auction is the governmental counterpart. But privatization has proven economically viable in numerous contexts (Walker, 1980, 1988). Thus the presumption ought to be in favor of this process as a means of promoting economic welfare.

MARKET IMPERFECTIONS

A major difficulty with Stroup's argument for government auctions of virgin land (as opposed to market-oriented homesteading)[10] is that it implicitly assumes the operation of a perfectly efficient public sector. The argument is that the market is imperfect; therefore, government intervention is justified. One may of course stipulate that if civil servants were omniscient and benevolent, and participants in the marketplace were ignorant, venal, and self-interested, then there would indeed be a strong economic case for supplanting the latter by the former. But the teachings of the Public Choice School, for one, are incompatible with any such presupposition. For it is a basic insight of such authors as Buchanan and Tullock (1971) that people do not suddenly sprout wings when they leave the private sector and enter the civil service. On the contrary, those who enter the public sector from the private bring with them all their human fallibilities, including ignorance and selfishness.[11]

The objection to Stroup is not limited to the self-evident claim that state bureaucrats are imperfect. Given that the functioning of both the public and private sectors is flawed, the issue is, which is worse? Stroup contents himself with indicating drawbacks in the market (homesteading) system. But in order to reach the conclusion he seeks, it is incumbent upon him to show that the graft and corruption usually attendant upon government operation[12] would not entail even more of a welfare loss than a blemished market. Even if his attack on homesteading were correct, in other words, his conclusions would by no means follow. Given that his argument is correct, he is entitled to conclude from it that both methods of property distribution (auction, homesteading) are imperfect; unless and until he can show that the former is less imperfect than the latter, he cannot logically prefer one to the other.

PERFECT COMPETITION

This brings us to the basic difficulty with Stroup's theoretical work. It is that it is far too beholden to the unrealistic model of perfect competition. Our author has embraced the notion of full and perfect information with a vengeance, but it ill accords with the real world, to the detriment of the explanatory power of his model.

This is seen when we ask about the kind of world that would have to exist for his hypothesis to be useful. It would have to be one in which many people knew full well of business opportunities which might not arise for decades. How else can we explain his fears that in the absence of state-organized auctions, people would be camping out for years, if need be, in order to prematurely homestead land which would only become economically viable dozens of years later. This behavior is eerily reminiscent of what occurs when a professional football team announces the sale, at prices far below market values, of Super Bowl tickets. Fans "camp out" on line, with their sleeping bags, thermos bottles, and picnic boxes, days in advance of the announced sale.

But the real world is not often like that. Rarely is knowledge of the future given to an imperfect mankind. And the proof of this is that entrepreneurship is continually with us (Kirzner, 1972). Were there to be full information available to market participants, profits would be zero, and entrepreneurship would come to a complete grinding halt. This may be true is the never-never land of most introductory economic textbooks, but as a description of the world as we know it, and, presumably, the one to which our author intends his model to be relevant, this is obviously false.

As a matter of historical fact, it was Columbus[13] who was the only one[14] to venture to the specific area that he visited in 1492. The great bulk of the United States was settled by people who, when they finally reached their destinations, were few, far between, and isolated. Typically, the lands they claimed were not even economic goods: there was more and as good that was not claimed. [15]

This situation of land surplus in the nineteenth century is entirely consistent with the Austrian emphasis on entrepreneurship (Kirzner, 1972). Here, one or a few individuals, with more foresight than their fellows, take advantage of a state of affairs that only they can perceive. With regard to homesteading, only several individuals (compared with land availability) typically anticipate that a migration will result in entrepreneurial gains, or profits, for themselves.[16]

The scenario propounded by Stroup is completely at variance with actual historical experience. In his view, thousands and thousands of people were contending for a very limited acreage.

And on the rare occasions that reality resembles the Stroupian model, there is almost always the not so subtle hand of government interference with the marketplace at work. For example, consider the cases where people would line up in horses and wagons, and then begin the race to homestead land when the government starting gun was sounded. This is the model of homesteading popularized in novels and cinemas.[17] But why did the people have to wait until the starting gun went off? It was because government had hitherto prevented land settlement!

Alternatively, assume that government today announced that in ten years' time it would acknowledge ownership rights to what are now the national parks, or the radio and TV broadcast spectrum, if they were to be homesteaded in the interim. There is little doubt that chaos would ensue. People would avidly pursue these opportunities, while their value would be dissipated in the process. But this would be due to the fact that such an announcement would have been made two centuries too late in the case of the land, and almost one century too late in the case of the airwaves.

PRIVATIZATION?

Another problem with leaving the dispersal of the virgin territory to the government is that instead of privatizing it, as Stroup implies they will, the state may instead keep the land under its own control,[18] or refuse to allow any ownership of it at all.[19] Under homesteading, however imperfect, at least we can count upon the resources passing into private hands.

In some cases, there may be good and sufficient reason[20] for government not to follow the implicit Stroupian advice and engage in massive property auctions. For example, there are no governments which now claim sovereignty over the oceans, seas, and Antarctica. Were any adventurous souls interested in economically developing these swatches of real estate, they could have no recourse to any nations as presently constituted. One option, of course, would be homesteading. The difficulty here is in ascertaining whether or not ones' property rights would be protected in future. Another option, this one more compatible with the public policy implications of Stroup's analysis, would be to have a super powerful United Nations, that is, a World Government, auction these properties, lest, horrors, any business firm engage in "premature" homesteading.

But the very idea appears almost ludicrous. It is painfully obvious that if even a government such as that of the United States—supposedly dedicated to the free enterprise system—would forebear to auction off its vast land holdings[21] to private owners, there is very little hope indeed that the nations of the world as a whole—most of which have fallen under the domination of abhorrent Marxist philosophies—would do so.

And what of the properties on the moon, Mars, and even the asteroids? Although perhaps premature as of the present, these resources may well become economically viable during the century which is now almost upon us. Would Stroup give over these resources to the tender mercies of a World Government—or a Solar System Government, should the discovery of other intelligent species lead us in that direction? It is one thing to entrust such entities with the protection of private property rights. This, even in the United States with an activist Supreme Court, is risky enough. It is quite another matter, as per Stroup, to ask such entities to first auction off land, and then to uphold the privatization.

As well, there is the difficulty that government, too, oft-times engages in activity which fits the Stroup model. That is, the public sector can also act prematurely. If we interpret the landing of three U.S. taxpayers on the moon in the Kennedy Administration as the first step toward colonizing the heavenly bodies, then it would appear that this was a very precipitate step. It was far in advance of what might have been called for in a more rational plan, which would have made sure that the basic technology, the development of complementary factors, and so on, was first in place before undertaking such a step. But if both the public and private sectors can act over hastily, then surely one does not prove the former preferable merely by pointing to this behavior in the latter.

Perhaps Stroup's perfectly competitive equilibrium[22] model better fits other property which has not yet been homesteaded, but may, one day, undergo such a process.

Perhaps the settlement of the American west is too onerous an historical episode against which to test its contention. After all, most people were completely unaware of the resource opportunities available to them through homesteading.

Instead let us consider the moon.[23] There can be no question but that everyone is fully aware of the existence of the moon. The U.S. moon launch in the past decade was widely publicized, so there can be no doubt that the knowledge that it is now technically possible to visit this worldlet is now widely disseminated.

Based on Stroup's analysis, we are not yet at the appropriate point with regard to moon settlement. As I write the cost of commercial lunar settlement would now appear to be prohibitive. That is to say were such activity to be begun at present, it might easily dissipate the present value of all future rents to be earned through settlement.

SUPER BOWL TICKETS

Stroup attempts to show the economic inefficiency of homesteading nineteenth-century land by likening this process to the sale of underpriced Super Bowl tickets on a first-come first-served basis in the modern era (p. 71). In both cases, he contends, below market pricing takes place. Let us take this analogy to its logical conclusion, however. Stroup advocates government intervention in the former case (a government auction of land is preferable to inefficient free enterprise homesteading), but he fails to carry through consistently in the latter. If he did, he would have called for government price controls on National Football League championship tickets, for their pricing policy—just as in the case of homesteading—creates a queue whose members pay "partly in the form of inconvenience and time lost" (p. 71). Perhaps by shedding light on the fallacies involved in urging government action in the football case (price controls) we may more easily appreciate the misconceptions which underlie the call for government action in the case of converting virgin land into private property (auctions).

It is easy to show that price controls—in all cases, and no matter how well intentioned their proponents—reduce the economic well-being of at least one of the parties to the trade. If this were not so, then, presumably, there would not have been any "need" for outside interference in the first place. If both parties can gain from a change in the terms of trade, the presumption is that the marketplace will gravitate to that state of affairs. In addition, the problems of price controls—inflexibility, allocative inefficiencies—are sufficiently well known in the profession. [24]

In the specific case of the Super Bowl, prices are indeed set far below the level which might equate the supply, limited by stadium capacity, and the now suddenly voracious demand. But this is not due to any inefficiency on the part of the sports team, as Stroup implies. Rather, ticket prices are kept purposefully low for the same reason that the grocer does not sell milk at market-clearing prices during an emergency (fire, flood, storm), nor the hardware store flashlights and shovels. All of these merchants, the grocer, hardware supplier—and the football team as well—have to deal with their customers when the abnormal demand comes to an end. If they raise their prices to market-clearing levels during these extraordinary times, the general public, unsophisticated in basic neoclassical economics, will resent them.[25] The average consumer will feel victimized by such a peak load pricing policy.

Entrepreneurs engage in short-run profit maximizing behavior only at a great risk to their entire operation (see Block, 1976).

It may be objected that (minimum) price controls would in this particular case, and perhaps in no other, have a positive benefit: if the football team (the grocer, etc.) is forced to raise prices to market-clearing levels, so that no queues are allowed to develop, the general public will not blame it; instead, such action will call down obloquy upon the head of the government board of price control. But this is unlikely. The state habitually and successfully blames the market for its own excesses.[26] It would be easy for detractors of the free enterprise system to assert that government was only imposing price controls in this case because of the previous "recalcitrant" behavior of the sports teams.

Even, however, on the assumption that state activity might in this case "work," and reduce the queues of avid football watchers, there are still difficulties. For one thing, it is problematic to equate a market-clearing price with the price that would arise under a free enterprise system. Yes, in a free market, there is a tendency for supply to equal demand, for there to be no queues, but this is only a (usual) result, not a defining characteristic of the system. We must distinguish between queues that are the result of government intervention and those that are part of the voluntary exchange system. It is clear that a queue due to a price ceiling is inefficient; at a higher price, willing buyers and sellers could have engaged in mutually satisfactory economic activity. But a private queue is a completely different matter. Here, it is patently *not* true that there are two parties to an exchange, *both* of whom are balked in their intentions. On the contrary, one of them (the NFL team, the grocer) refuses, let us suppose that this is for reasons best known to him, to engage in a trade at a price that would equate amount supplied and demanded. Forcing him to do so by coercive means would thus clearly reduce at least his welfare.

OTHER "GOODS"

But land is hardly the only commodity which can be described by Stroup's figure 1. There are many other goods, some far more valuable than acreage, which initially yield a negative return, eventually pass through a break-even point, and then finally garner a positive dividend for the investor.

Human relationships, particularly romantic sexual liaisons, also follow this pattern. Moreover, at least at the margin, the utility derived from such interpersonal relationships are often far more important than land. Helen of Troy launched a 1,000 battleships; acreage in Athens was never even in the running for this honor. Men come to blows, even murder, over the favors of a beautiful woman; true, land is also the object of warlike behavior, but when this occurs it is typically less passionate, personal, and bitter.

If the Stroup hypothesis were correct, one would expect it to apply to the "marriage market" as well as to land. That is, one would expect it to be common behavior for adults to try to "homestead" young children who were particularly attractive, or who showed promise of developing into superlative candidates in the "marriage market."[27] An adult could buy lollipops for the young person, take him (or her) to the zoo, and so on, in an attempt to ingratiate himself (herself) with the child.[28]

Yet the reality is, to say the least, very different from this model. The only adults who hang around children in such manner are the perverts and child molesters whose interest is very much with the *present*, while their "homesteadees" are still very young.[29] They are not at all concerned with the future, when these children will have achieved adult status, and thus be no longer attractive to them. Indeed, it smacks heavily of the ludicrous to even entertain the notion of healthy adults queuing up in line to await the maturation of children in this way. Nor is this somewhat a bizarre notion obviated by any analogy to a government auction for young children.

RESEARCH AND DEVELOPMENT

But all this is highly speculative. We do not have to resort to such fanciful counterexamples in order to refute the Stroup hypothesis. Were it true, other more mundane results would occur, which do not actually take place. Consider scientific and technological research and development, for example. This, too, is a direct analog to the homesteading of virgin land. Just as in that case, and initial investment is made, in the hope that it will bear fruits in the long run. Assume, then, that the R&D investor faces a horizon depicted as in Stroup's figure 1. The ideal time period for him to make his discovery would be at t*. But, just as in Stroup's analysis of the homesteader, our investor may fear that others will steal a march on him, beat him to the punch, and discover the new process before he does. So, he will be led by the invisible hand of competition to begin undertaking his investment long before the optimal point t*. He will start, just as his homesteading counterpart, at t. Why? "An individual whose view coincided with Figure 1 would be willing to (begin research) as early as time t, calculated to be the time at which the negative value of the time spent between t and t* would just equal, in present value, the sum of future positive values to be captured by the (investor). If he believed other (competing investors) to have similar views, he might indeed (begin research) at time t, thus destroying by premature (investment) an amount of wealth equal to all the present value of future rents believed to exist in (discovery of the new invention)."[30]

If Stroup is correct, then, in advocating government land auctions instead of private homesteading, on the ground of wealth maximization, he must also consent, and for the same reasons, to an analogous public-sector "sale" of the benefits of scientific R&D. For only in this way may the economic horrors of premature research investment be precluded.

Can Stroup argue that there is a disanalogy between land and technological development? That on the one hand the land to be homesteaded is well known to all and sundry, but that on the other hand the results of research are all but unknown at the outset, and that further there is great risk that any given experiments will fail?

Were this reply to be attempted, it would only expose the argument, all the more, to the charge that it is truly mired in the perfectly competitive assumption of full information for land. But it is simply not true that all potential homesteaders know full well of all opportunities open to them; that homesteading is a riskless exercise.[31] On the contrary, entrepreneurship, as we have seen, pervades every nook and cranny of the real world economy. Yes, it pertains to research and development, but it also applies to the settling of new lands.

A further distinction needs to be made. It is undoubtedly true that there will be less risk with regard to the *existence* of lands potentially to be homesteaded than with regard to their future economic viability. But only the latter is really relevant to the decision to engage in homesteading. And the latter is, on a par with R&D, fully susceptible to risk and uncertainty.

CONCLUSION

I must conclude that despite presenting a dramatic challenge to the institution of homesteading, Professor Stroup fails in his attempt to undermine it. There may have been a case of overoptimistic homesteading in the 1860s; if so, our author fails to demonstrate it. Nor does he show how "economic logic" (p. 76) suggests that people would prematurely settle virgin lands. His the-market-is-imperfect-therefore-government-action-is-justified argument has a long pedigree, and Stroup deserves every credit for adding to it in a creative manner. I venture to think that his challenge is easily as profound as the monopoly, externalities, or depressions-are-intrinsic-to-the-market arguments. But like them, ultimately it cannot rigorously be maintained.

NOTES

1. The author would like to thank the following for helpful comments: David Friedman, Gordon Tullock, Terry Anderson, W.A. Albada and, somewhat paradoxically, Richard Stroup.

2. All page references, unless otherwise indicated, refer to Richard L. Stroup "Buying Misery with Federal Land," *Public Choice,* Vol. 57, 1988, pp. 69–77.

3. For a defense of this view, see Rothbard (1982); for a critique, see Nozick (1974).

4. The plight of the European immigrants who landed in New York City was great; far worse was that of those who suffered from the Irish potato famine; even the gold miners on the U.S. Pacific Coast in mid-nineteenth century are said to have led economically unsatisfactory lives in the extreme. One can hardly deduce Stroup's case of resource misallocation or lost land rents from mere poverty; this state of affairs, unfortunately, was widespread.

5. U.S. Department of Commerce (1975, p. 912).

6. To be sure, they are very rough figures. They well might not be finely calibrated enough to shed light on the Stroup hypothesis. For one thing, they apply to the United States as a whole; they do not distinguish between state subject to the Homestead Act, and those which were not. For another, they apply to business failures, not personal bankruptcies (I owe this point to David Friedman), and homesteader losses would be heavily weighted toward the latter. But the burden of proof should be placed on the counterintuitive premature settlement scenario; not upon its denial. As well, Stroup offers no evidence showing that the personal bankruptcy rates in the states subject to the Homesteading Act were significantly higher than those where this law was inapplicable. As a result, I cannot maintain that I have empirically demonstrated that there was no great suffering; only that Stroup's thesis remains unverified, despite the evidence he presents.

7. Ironically, this view is also expressed in Baden and Stroup (1988). As well, see Libecap and Johnson (1979).

8. Strictly speaking, there was no civil war in the United States in 1861–1865. This term properly implies to fighting between two groups, wherein each of which wishes to rule the entire country. For example, there was a civil war in Spain in 1936 between the fascists and the communists. There was another starting in 1917 in Russia between the Red and White armies. But in the occurrence in 1861 while the north did indeed desire to govern both areas, the south did not want to do so regarding the north. In contrast, it wanted to secede from that jurisdiction. So this was a war of secession, not a civil war.

9. Stroup attempts to justify it on the grounds that it is better that the "federal treasury absorb" the resources frittered away on premature settlement rather than that there be "no offsetting economic benefit to the nation" (pp. 71–72). But this is a pure value judgment, which cannot be supported by anything in the realm of positive economics. One might with equal justification claim that it is better that resources be wasted, if wasted they must be, then that they be given over to the evil government, which will use them to further solidify its pernicious hold over the economy. This, too, would be value judgment, equally incompatible with a positive economic outlook.

10. See Anderson and Hill (1983), Rothbard (1982).

11. According to Hayek (1944) the situation is even worse, as the system of dirigisme encourages the "worse to get on top."

12. We must resist the objection that the crookedness that usually plagues government operation would not apply in this case because here the state is merely selling off virgin lands, and is not engaged in production or commerce on its own account. On the contrary, government corruption is endemic, too, in cases where it is "merely" attempting to contract out, to the private sector, activity formerly in its domain. The recent (1986) bribery scandal in the New York City Parking Violations Bureau is a case in point. Geoffrey G. Lindenauer, the deputy director of this organization pleaded guilty to racketeering and mail fraud for soliciting bribes from private firms who sought contracts with his agency. There is little reason not to suppose that this sort of graft would be and was an intrinsic part of government operation and management of land sales through auction. For the view that such activity is intrinsic to the public sector, and central planning of whatever stripe or variety, see Hayek (1944).

13. This applies to Magellan, Cortez, Pizarro and other such explorers as well.

14. We ignore the possibility that the Vikings may have visited the shores several centuries beforehand. They did not stay to settle and homestead, in any case.

15. This is in sharp and dramatic contrast to Stroup's discussion of the "winner's curse" (pp. 73–75) where he envisions as typical the situation where there are numerous potential homesteaders all seemingly poised at the starting blocks, ready for the timer's gun to go off signaling the beginning of the race.

16. Economic history is compatible with the view that these settlements were economically viable at the time they were settled. Thus, in our model they arrived after point t*. To be sure, many of the settlers subsequently picked up stakes and moved further west. But this is not inconsistent with the view that they first took up residence at or after point t*, for two reasons. First, others took over their leavings and were able to earn revenues above their alternative costs, as indicated by their long-term occupancy. Second, the first arrivals may have been economically secure, but may have seen even greener pastures further west.

17. And based on the experience of the Sooners ("the sooner you got there, the better") of Oklahoma.

18. In Alaska, land in government hands comprises 98 percent of the total; only 2 percent is privately owned.

19. A situation of "non-ownership" best describes the state of affairs with regard to the oceans, lakes, rivers and streams, and the heavenly bodies.

20. Or at least what some people might consider good and sufficient reason.

21. Many of which are, as I write in September 1988, in the process of being consumed by fire. Some commentators have blamed the fires now raging in Montana, Idaho, and Wyoming on the "free burning" policies of the radical ecologists who infest the U.S. Forest Service. But a more basic criticism is to be leveled at the principle of public ownership in the first place. After all, a private owner who did not protect his resources from such "natural disasters" would soon enough be forced into bankruptcy. It is only under the institution of public ownership that such erroneous philosophies can do such stupendous harm.

22. David Friedman (1973) has proposed a similar claim regarding the booty available as a result of engaging in political activity. He theorizes that the earnings from such behavior will be no higher than those which can be obtained through

other lines of endeavor, and for similar reasons: the resources devoted to gaining political pull will, in equilibrium, just offset the revenues derived thereby. This is obviously true at the formalistic level: this is a correct description of the equilibrium state in which there is no incentive for anyone to change. The relevant question is, however, does the economy ever reach such a point? That it does is counterintuitive in the extreme. It implies, among other things, that all participants in the political process are equally able in perceiving new rent-seeking opportunities; that there always exists a perfect (or at least near-perfect allocation of resources in society. Were these conditions to hold, there would be no reward for entrepreneurship. But as we know that there is the contention of full resource allocation must be false. If illegal activities such as those undertaken by the Mafia are interpreted as a sort of rent-seeking (and there is no reason why they should not be) the implication of the Friedman-Stroup hypothesis is that this organization never earns super normal profits. Where, then, do they get all that money for limousines? Is the Godfather a complete economic fiction? Are we to leave all our illusions broken by the perfectly competitive model?

23. We cannot, in this regard, realistically use the oceans or the Antarctic for this purpose. These entities cannot be privately homesteaded because no government stands ready to defend any such settlement and commercial exploitation. The situation with regard to the moon is perhaps equally unclear, if not more so. Nevertheless, for the purpose of discussion, we shall assume there to be no legal barrier homesteading the moon.

24. Frey, et. al. report that only 5.7 percent of U.S. economists, and only 2.2 percent of their West German counterparts agreed with the statement "Wage-price controls should be used to control inflation."

25. Alternatively, one might interpret the low ticket prices as gift on the part of the NFL team to its loyal customers. A minimum price control would render this gift impossible, and thus make both parties to the transaction worse off. See Kahneman, Knetsch and Thaler (1986).

26. An example that springs to mind in this regard concerns inflation. Government, and its defenders, commonly attribute inflation to private "greed," and "self-seeking"—whether on the part of consumers who buy too much, retailers who "profiteer," or union leaders who are "grasping." But as is well known by now in the profession, it is government's own (central banking) money creation which is the cause of this particular malady.

27. It must be readily admitted that this is only an imperfect analogy. In the case of land, one is virtually assured of actually receiving that which one homesteads; in the case of children, one can do no more than increase the probability that one's "suit" will be looked upon with favor when adulthood reached. Land is inert; it must "accept" its homesteader. People make choices; thus, they may decide to reject their "homesteaders."

But this distinction can easily be overdrawn. Nothing in the future is for certain in this world. There is no *guarantee* that the homesteader of land will actually receive it. Stroup (p. 71) reports that "hardly more than a third of the homesteaders (of land) before 1890" actually succeeded in their quest.

28. Arranged marriages, such as occur in India, are not relevant to our concerns. These are devised by two sets of parents for similarly aged children, usually at birth. What we are here considering is the case of adults queuing up in line, waiting for children to grow up, prematurely anticipating them, as it were.

29. One searches in vain for real-world counterexamples; literary ones are hard to come by as well. However, one such does exist: Robert Heinlein (1973). I owe this point to David Friedman

30. Paraphrase of Stroup (1988, pp. 73–74). Material in brackets inserted by present author.

31. Stroup's own comments about the economic hardship suffered by nineteenth-century settlers beam eloquent testimony on this matter.

Chapter 7

Property in Space

Space Environmentalism, Property Rights, and the Law

INTRODUCTION

A considerable body of academic literature exists on the subject of space law, despite the fact that very few human beings have ever been to outer space, and substantial presence beyond the earth's orbit still seems to be in the rather distant future.[1] Among those proposing detailed, centralized regulatory regimes for a realm that so far has little, if any, need for them are anti-market[2] environmentalists who, not content to attend merely to the earth's pollution problems, are concerned about the possible future pollution of the moon and other celestial bodies such as Mars. This chapter considers the ideas of these individuals—whom we will call socialist space environmentalists[3]—and rejects them.

We find proposed environmental programs for outer space not only philosophically ill-founded but also economically and pragmatically unjustified. In their place, we propose an alternative: a regime of pure private property and strict liability. Though our approach might be considered radical if proposed for the earth,[4] we maintain that it is an entirely appropriate policy for outer space and celestial bodies.

Types of Environmentalism

Few, if any, human beings would self-identify as enemies of "the environment." After all, everyone wants clean air to breathe and clean water to drink, and does not want anyone to invade his person or property with harmful substances without permission. People who go this far—and only this far—with their environmentalism probably comprise the majority of humanity. They can be said to be adherents of *anthropocentric* environmentalism (Flournoy, 2003). Anthropocentric environmentalists can be found across the political spectrum. For example, voices ranging from the right (Posner, 2003) to the extreme Marxist left (Commoner, 1990; Ehrlich and Ehrlich, 1968, 1981; Gore, 1992) have called for unprecedented global government intervention to combat perceived environmental threats to human well-being. Others, however, have advocated laissez faire capitalism as the appropriate means to protect the environment to maximize human well-being on Earth (De Alessi, 1991; Hill and Meiners, 1998; Machan, 1984). For the anthropocentric environmentalist, nonhuman creatures and objects are valuable to the extent that humans value them—they have no "intrinsic" value apart from this.[5]

In the second half of the twentieth century, another type of environmentalism came to the fore: *ecocentric* environmentalism. *Ecocentric environmentalism* is also sometimes referred as "deep ecology," to contrast it with the "shallowness" of anthropocentric concern for the environment (Scriven, 1997). Originated by Aldo Leopold, who conceived the idea of the "land ethic," ecocentric environmentalism holds that the environment itself is intrinsically valuable and that human beings themselves have value only to the extent that they play a role in, and support, this environmental whole.[6] According to radical ecocentrism, only "ecological wholes (such as species, ecosystems, the land or the biotic community) . . . have a value in themselves . . . and . . . the value of the ecological parts . . . is determined by how far they contribute to the survival and well-being of the ecological whole."[7] The ecocentric view is not limited to concern for animals or even plants, but to the entire Earth, dirt and rocks included.[8] Everything on Earth, except for humans, is seen as possessing "intrinsic value" (i.e., value somehow derived from itself, not from man) which is destroyed or threatened by any human tampering at all.[9]

The real-world implications of this philosophy can be seen, for example, in the activities of the Earth First! organization, which is known for, among other things, putting spikes in trees so lumberjacks or mill workers who cut them may be injured or killed.[10] Earth First! leader Richard Foreman states the ends of ecocentric environmentalism bluntly: "We advocate bio-diversity for bio-diversity's sake. That says man is no more important than any other species. . . . It may well take our extinction to set things straight."[11]

To be fair, not all ecocentric environmentalists go this far. Advocates of what could be called "strong" (but not "radical") ecocentric environmentalism value both the environmental whole (independently of its value to humans) and human beings as individuals.[12] "Weak" ecocentric environmentalists similarly value both, but ultimately are willing to put humans first.[13]

Ecocentric environmentalists typically speak in terms of the Earth, "biodiversity," and the "biosphere."[14] Indeed, radical environmentalists who subscribe to the "GAIA hypothesis" consider the Earth itself to be a "superorganism" and see the Earth goddess Gaia (or "Mother Earth") as "an embodiment and distinctly a personification of the intrinsic value of the Earth."[15]

Considering this strong focus on the Earth, one might expect that we would be spared the down-with-humans-up-with-dirt-and-rocks rhetoric with respect to man's activity beyond the earth. Many ecocentrists, however, appear ready to expand their area of concern to any place man might travel. Thus, as we are arguably poised to move significantly into outer space and other celestial bodies in the next few decades, we already have calls for "cosmo-centric" environmentalism[16] or "astroenvironmentalism."[17] Howard A. Baker writes approvingly, "With an environmental approach, protection of the outer space environment and its sub-systems is the priority," not "ensuring that outer space can be used for [human] space activities."[18] "Outer space, a source of wonder and inspiration for centuries, deserves to be preserved *in its original pristine state, for its own sake* and for future generations to enjoy," writes Bernard K. Schafer.[19] Another writer states, "[W]e must ensure that our presence [in space] does not defile what remains one of the few accessible pristine areas."[20]

Ryder W. Miller lists among space environmentalism's goals the declaration of "celestial bodies [as] pristine wildernesses that need to be protected rather than frontiers to conquer"; prohibition of "terraforming" of celestial bodies (e.g., attempting to transform Mars into an earthlike planet where humans could live); "safeguarding against the introduction of non-terrestrial life to and from celestial bodies"; and prohibiting all property ownership by both governments and private parties[21] in space.[22]

Space environmentalists Paul F. Uhlir and William P. Bishop, quoting Harry H. Almond, Jr., suggest five principles to guide space environmental policy: (1) "environmental balance" (with "substantial weight given to intrinsic environmental values"); (2) conservation of resources ("The natural resources of outer space must not be exploited in a wasteful or environmentally damaging way."); (3) *Absolute liability for wrongful or negligent acts to the space wilderness and the duty to restore conditions to the status quo ante*;[23] (4) demilitarization; and (5) permission for "states" to use and explore space, provided they abide by the first principle of "environmental balance."[24] They also call for zoning laws for celestial bodies to restrict the uses to which land can be put.[25]

Even relatively moderate, generally anthropocentric observers appear to have been influenced by the ecocentrists' views and have called for preservation of pristine "wilderness" areas on celestial bodies. Glenn H. Reynolds and Robert P. Merges, for example, generally favor private property rights but make an exception for "environmental research and conservation preserves," which would place "10 to 15 percent of the area capable of being developed" off-limits.[26] They offer no argument to support this notion but instead state, "It should not be necessary at this point to defend such an idea."[27] Similarly, Lawrence D. Roberts calls for "[r]estrictions on development which gradually open the high frontier over time" on inhospitable bodies, and a "precautionary principle" restricting development where life might have once existed, presently exists, or could potentially exist.[28] Others with no otherwise-apparent hatred for the human race make passing reference to the necessity of some degree of space environmentalism.[29]

We do not intend to attack the radical ecocentric environmentalists directly in this chapter. The absurdity of "intrinsic" value or value independent of the existence of human beings has been well-refuted elsewhere,[30] so we need not rehash those arguments here. Besides, we assume rather reasonably that the majority of our readers are anthropocentrists or at least moderate ecocentrists who do not favor the human race's demise over and above any disturbance of the rocks of the solar system. And, of course, there is little use in trying to rationally persuade adherents of a religion, including nature worshipers, with rational arguments—they will believe what they will believe regardless of what we may write here. Further, their position amounts to a logical or internal contradiction. They wax eloquently about overpopulation.[31] Yet they all, each and every one of them, have the power to reduce the number of the earth's inhabitants by precisely one. The fact that they are still here, complaining bitterly of too many people utilizing too many resources, shows that they do not take their own views seriously. If they do not, why should we? These people might likely argue (not totally unreasonably, given their premises) that they reduce the population more by sticking around and persuading others not to have children, and so on. However, if this were true, then, when they were no longer able to convince people of the merits of their position (say, due to old age or infirmity), they would commit suicide. They might well do so publicly, in order to better promote the overpopulationist movement. To the knowledge of the authors of the present chapter, this has *never* been done. Actions speak louder than words.

Instead, we intend to argue to our fellow anthropocentrists and even moderate ecocentrists that one would have to be a *radical* ecocentrist to seriously embrace any sort of space environmentalism that goes beyond a regime of full private property rights and strict liability for harms to private property. There is simply no legitimate philosophical or pragmatic argument to the contrary. Nor, we will show, does international law demand anything more than this.

What Is the Space Environment?

To speak of a "pristine" environment outside of the planet Earth is a rather strange thing to do, given how utterly unpleasant the rest of the solar system (and, as far as anyone knows, the universe) is. The planet Mercury, for example, has no atmosphere, and portions of its surface become hot enough to melt tin. Other parts, however, remain cold enough to keep ice from crashed comets perpetually frozen—and there is nothing remotely pleasant in between. Mercury is, in one writer's words, "geologically dead. It has not changed significantly in several *billion* years."[32]

Venus is even worse—"a good substitute for Hell."[33] Its atmosphere is a "choking shroud of almost pure carbon dioxide" (a gas much hated by environmentalists on earth) complemented by "thick clouds of battery acid."[34] Its atmospheric pressure is ninety-two times that which exists on the earth's surface, so any visiting astronaut in a spacesuit would be "crushed instantly."[35] And the mean surface temperature is 480 degrees Celsius—even hotter than Mercury, and hot enough "to melt tin, zinc, and lead."[36]

Earth's moon is relatively less hateful, but it has no atmosphere, of course, and "has never supported liquid water," let alone life.[37]

"Mars is not alive. It is dead, and looks as if it has been that way for a long time. No conclusive evidence for life there, either now or in the past, has ever been found."[38] Its atmosphere consists mostly of deadly carbon dioxide,[39] and its mean surface temperature is −23° degrees Celsius.[40]

The planets further out are even worse, so bad that it is difficult to imagine that they could be of any use at all to humans, except perhaps as something for tourists to fly past and admire. Jupiter, Saturn, Uranus, and Neptune are covered in extremely cold, giant, stormy mixes of toxic liquids and gasses.[41] Tiny Pluto apparently no longer counts as a planet, and has a surface temperature of −230 degrees Celsius and an atmosphere of nitrogen (good) and methane (poison).[42] There is talk of a tenth planet, but we are not competent to pronounce on its status.[43] Some of these distant planets' moons might be of some use to humans but are nonetheless wholly inhospitable. For example, one of Jupiter's moons, Europa, is covered in water ice and may have liquid water and possibly some sort of microscopic life beneath its frozen surface. And Saturn's moon Titan has, like Earth, a mostly nitrogen atmosphere—at −180 degrees.[44]

Where there is no atmosphere, as on the moon, the environment is far from healthy. Spaceships and spacesuits must be well shielded to protect against the sun's radiation. "A hypothetically unprotected astronaut would receive (in the absence of solar flares) about 10 rems of radiation per year. In comparison, the average person on the face of the earth receives only about .1 rems of radiation per year from background sources (e.g., from the earth and from space)."[45] The presence of solar flares makes matters much worse—their high-energy protons "can cause the release of lethal doses of secondary radiation, such as gamma rays, when they collide with spacecraft."[46]

All of that may sound bad, but in fact the space environment is only going to become worse, much worse, even if we humans never reach other celestial bodies. That is because, as the eons pass, our sun will eventually change to a "subgiant" star, then a Red Giant, then a nebula, then a White Dwarf, then a Black Dwarf. In the end, all of the planets, including Earth, will lose their atmospheres and exist at a temperature just a few degrees above absolute zero, the coldest temperature possible.[47]

Thus, in sum, the space environment is so bad right now that, from anything other than a rock-and-dirt-worshiping perspective, it could not get much worse—except that billions of years from now, it will get worse, and there is nothing anyone can do about that.[48]

PROPERTY RIGHTS AND PROBLEMS OF SPACE POLLUTION

Is Space Pollution Even Possible?

Considering the solar system's present and future environmental state, the idea of space pollution becomes absurd to anyone apart from those who believe that whatever is "natural" is best, including even the eventual frozen desolation of our solar system. The reality is, virtually nothing human beings could do to the solar system could likely make it less livable or less useful than it is now.

Air pollution? As we have seen, there is no air on the moon—and to the extent that our neighboring planets have an atmosphere at all, it is almost entirely carbon dioxide, which is toxic and the bane of environmentalists here on Earth when it is produced by our automobiles.[49] Thus, nothing we could do to other celestial bodies could make the "air" more toxic than it already is. Water pollution? There is no surface liquid water, anywhere but Earth. Radiological pollution? As we have seen, there already is dangerous radiation in space against which humans must shield themselves. The Mars atmosphere may limit the amount of radiation on its surface—but if one cannot live there anyway without special protection (given the poison-gas environment), just how much worse would some radiation here and there make the planet? Also, Martian soil is believed to be highly toxic, to the extent that it could even threaten completion of any manned mission there.[50] Any human vehicles or structures there would have to be specially protected by something like "mega-Rust-Oleum" to avoid destruction from this planet's violent sandstorms.[51]

Thus, to speak of pollution or contamination of space in the abstract—apart from human beings' property rights—makes no sense.

Air and Water Pollution in Space

Lawrence D. Roberts suggests that "[u]biquitous commons resources on Earth such as air and water will likely pose the same kinds of environmental challenges for space developers as they do for Earth developers," adding, "The need to recycle such valuable commodities will require stringent regulation of the discharge of hazardous byproducts into the waste stream."[52] We find this implausible. To the extent that there would be any air or water on the moon or elsewhere in space, how would it get there? It could only be from humans who brought or created it there. Where would it be found? Inside the space vehicles or other structures individuals brought to or built on the lunar (or Martian) surface. And here we get to the key of space environmental policy: to protect humans' environment in space, we need only protect their private property rights in whatever settlement they establish there.[53]

On Earth, such a policy, though philosophically sound, has historically presented some technical difficulties. For example, on our planet it may be difficult to determine which factories contributed to victims' air or water pollution and in what amounts, as contaminants may travel imperceptibly over long distances.[54] Pollution victims may also suffer very small harms individually such that a lawsuit would cost them more than it was worth.[55] Those problems are far from insurmountable in the earthbound context[56]—indeed, technological advances and the availability of class-action lawsuits should make them decreasingly problematic—but they do exist.

In space, however, apart perhaps from radiological poisoning, some sort of unmistakable physical invasion would be necessary for anyone to pollute anyone else's air or water. Thus, enforcement of a strict-liability regime for pollution should be simple and effective.

Other Pollution on the Moon and Celestial Bodies

Roberts also sees a need for environmental regulation on the moon to prevent pollution from lunar dust.[57] The extent to which this would be a problem requiring regulation is, however, unclear, given the moon's lack of an atmosphere. Further, given the moon's size and its likely sparse population even once humans begin exploiting it, it seems those using the moon for mining and those using it for recreational purposes or for a good view of the Earth would rationally spread themselves apart. With relatively few parties and a strong incentive to spread out, we can imagine that parties might bargain in advance to avoid conflicts, or later to eliminate them.[58]

Of course, to the extent that polluters (whether by dust, chemicals, radiation, or anything else) arrive at the moon first, they may establish property rights there *including the right to "pollute."* Where no one has already homesteaded lunar or planetary land, a mine or factory owner may homestead an easement to emit dust, and other potential pollutants, over the surrounding area that his operation affects.[59] Then new arrivals will know that they should not locate in the area the established industrial operation affects unless they are willing to subject themselves to the industry's by-products. Not only is this philosophically sound from a property-rights perspective,[60] it also should present little burden in practice because of the moon's likely sparseness for a very long time.

On the other hand, where the owners of hotels, golf courses, "wilderness" preserves, and the like arrive first, they will homestead their land, including the right not to be disturbed by pollution. Should someone trespass upon their property with any form of pollution, they will be entitled to both damages and injunctive relief, just as pollution victims were in Great Britain and the United States through the 1830s.[61]

Space: The Ultimate Waste Dump

One of the most promising uses for space is, of course, as a waste dump. This should be cause for environmentalist celebration, not alarm.

For example, rational observers recognize that nuclear electric power is far better for the environment than fossil fuels, which pollute the air and cause countless health problems for those of us who breathe in the pollution.[62] An important problem, of course, comes in the form of the radioactive waste produced, which, though small in quantity, remains hazardous for a very long time.[63] Once space flight becomes sufficiently affordable, the answer of what to do with this waste becomes simple: send it on a long, long trip.[64] Who but the most fanatical "cosmo-centrist" could be disturbed by sending all of our toxic waste to, for example, Venus, an already hellish place where no human being or other living creature will likely ever go? Even if we were to take the nuclear waste to someplace humans might want to go, such as the moon or Mars, the physical quantities of such waste are small enough that it would require only minimal space.[65] The only colorable objection to this is that the waste might pose a risk to people on Earth as it leaves the atmosphere (e.g., if the ship carrying it explodes or crashes, as NASA vehicles are wont to do[66]). But presumably that risk would ever decrease as the private sector moves further into the space transportation field and space technology advances. For example, a space elevator would not entail the high risks or costs of ordinary space flight.[67] And, of course, carriers of hazardous waste would be strictly liable for any harm—which, along with their financial investment, would encourage them to take extreme care.

Another potential benefit would be to move polluting industrial operations off-planet.[68] Again, environmentalists who really care about well-being of humans or life generally (as opposed to rocks and dirt per se) should delight in this prospect.

"Wilderness" Preserves

As noted above, Reynolds and Merges call for 10 to 15 percent of the moon (and presumably any other celestial body at which humans may someday arrive) as a "preserve."[69] They make no argument to support this view, however, and even explicitly state that they believe no such argument is necessary.[70] The radical environmentalists go further, of course, and want earthly governments to declare all of outer space an untouchable "wilderness."[71] We find the alleged need for official wilderness preserves less obvious.

Indeed, this seems rather a strange preoccupation. As we have noted already, there seems to be plenty of room up there for everyone and every purpose imaginable. Right now space is de facto 100 percent wilderness preserve, and it is difficult to imagine humanity making a significant dent in that number anytime soon. After all, even vast amounts of the relatively hospitable continents of North America and Australia have minimal population density, even excluding national parks and other areas governments have so far placed off-limits. Environmentalists have also purchased land for the purpose of keeping it vacant and preserved on Earth;[72] there is no reason they could not do so in space—or rather than purchase simply homestead because it is there for the taking.

Governments have little incentive or ability to determine which parts of any celestial body are best used as wilderness preserves or which are best put to other purposes.[73] One can imagine that such determinations may be corrupted by the inevitable influence of special interests, just as special interests have influenced terrestrial environmental laws to the benefit of polluters.[74] Indeed, the U.S. government's management of its national parks has been dismal, as have the environmental records of governments, especially socialist governments, generally.[75] Thus, if optimal preservation of that which is valuable to scientists and other admirers of pristine lunar wilderness is the goal, the answer again is strictly enforced private property rights.

It is entirely unjust for "wilderness" advocates to use government to prevent others from developing their property in space. As Glenn Reynolds has noted, theirs is essentially an "aesthetic view masquerading as a religious one."[76] They may speak in terms of intrinsic value, but they really seek to use the law to forcibly place their personal aesthetic preferences—their own human desires—above those of others, and above the welfare of the human race. By and large, they have been allowed to do this on earth, albeit only partially. Unfortunately, there is no reason why space should be any different. As we have seen, however, in space there is even less cause to cater to their desires. Perhaps, then, they will not succeed as well in the heavens as they have on earth.

Terraforming

"Terraforming" would involve transforming an alien environment (Mars being the obvious candidate), to give it a climate more like Earth's.[77] Fantastic though it sounds, this is likely to be technologically feasible on the fourth planet.[78] Essentially, it would involve initiating "global warming" through the release of carbon tetrafluoride (CF4) into the now very sparse Martian atmosphere, raising the temperature by 10 degrees Celsius within several decades, which would cause an increase of water vapor in the atmosphere, further warming the planet.[79] With that accomplished, humans could release "methanogenic and ammonia-creating bacteria into the now-livable environment,"[80] creating even more greenhouse gases. "The net result of such a program could be the creation of a Mars with acceptable atmospheric pressure and temperature, and liquid water on its surface within fifty years of the start of the program."[81] Mars would not then have a breathable atmosphere, "but would support crops and allow people to move around without spacesuits."[82]

Of course, those who want to preserve the outer space environment in its "pristine" form can be expected to oppose this.[83] Some already have.[84] From an anthropocentric environmentalist perspective, however, we find no problem with terraforming.

If no one owned property on Mars before terraforming apart from the terraformers, property rights certainly would not be an issue—the terraformers would have a right to do as they please with presumably large parts of the planet.

That is, terraformers could make whatever use they see fit or as much of Mars as they could homestead before other homesteaders arrived, apart from the land already taken over to use as a base for their operations. Strictly speaking, they could not claim even 1 square inch of the rest of the surface, as, by assumption, they would not have directly "mixed their labor with"[85] any of it. In the scenario we are positing, their actions were limited to setting off gases. Of course they would have a tremendous advantage in terms of homesteading over any other later arrivals.[86]

If other property owners did exist, they would of course likely welcome the change because it would make their own property more useful to them. (In economic terms, they would enjoy *positive* externalities of the terraformers' activity. Instead of "coming to the nuisance," they would be coming to its very opposite.) But some Mars property owners—especially scientists trying to research the planet's history—might not welcome the radical changes to the planet. Still, the right to be protected against weather one finds undesirable has never been recognized, to our knowledge.[87]

On the other hand, the very notion of externalities, whether positive or negative, is highly problematic. This concept is very subjective:[88] "One man's meat is another man's poison." A home painted with psychedelic colors will be off-putting to most people, but attractive to some. To take a case from our present concern, the search for extraterrestrial intelligence (SETI) is seen by many as a progressive step;[89] an attempt to uncover other species in the galaxy with whom we humans can interact. However, an entirely different interpretation of this initiative could be that it is very dangerous, in that it might only call attention to ourselves with regard to a superior but evil species who will either enslave or eat us.

Of course, non-property-owning environmental activists on Earth—those most likely to challenge terraforming—would have no standing to challenge this process of development. Again, their aesthetic tastes should not be given priority over those with an actual stake in the matter (i.e., property owners) and over the good of the human race generally.

What if There's Life?

Some observers, such as Roberts, believe that bodies "with the potential for harboring biotic or prebiotic activity" present a special case for which different rules must apply. Roberts states that where life exists or even potentially exists, we must apply the "precautionary principle," which would place the burden of proof on those engaged in a "challenged activity" and prohibit development that threatens evidence of past life or the existence of present or "potential" life.[90] We disagree.

First, we note that there is no evidence that life exists or has ever existed anywhere in the solar system except Earth.[91] Further, there is a strong consensus that to the extent that life might exist or have ever existed elsewhere, such as on Mars or Europa, it is limited to extremely simple microscopic organisms.[92] The likelihood of sentient or even plant life existing elsewhere in the solar system appears to be zero, and the question of life on planets outside the solar system is very hypothetical, even for an article on space law.[93] Therefore a presumption against the existence of *actual* life where no evidence to the contrary exists seems proper.

Further, space environmentalists have failed to make the case that environmental regulations are necessary to protect whatever extraterrestrial life (or evidence thereof) may exist. Humans are fascinated by the prospect of the existence of any kind of extraterrestrial life. Anyone who bothers to go to space for any purpose is likely to be interested in checking for signs of past or present life on his property (or prospective property) before acting in a way that might destroy it. For the intellectually uncurious, there would still be financial incentives. For example, scientific or environmental organizations could offer prize money for the discovery of evidence of extraterrestrial life; a property owner who discovers evidence of life could sell scientists, journalists, and others rights to access, study, and publicize information about the discovery. Only governmental intervention (e.g., stripping individuals of property rights when something of scientific interest is found on their property) is likely to cause incentives to run in any other direction.[94]

Suppose there were the proverbial "little green creatures" discovered on Mars, or on any other planet humans colonized. What rights would they have? What obligations would we have to respect these rights? (If they were smarter/stronger than we, the shoe of course would be on the other foot.) There are several options. If they had the intelligence/ability of dogs or cats, then we would treat them as we now do those animals. But suppose they were intermediate between us and the smartest of earth animals (chimps, porpoises), or had human qualities but looked like a cross between an octopus and a giraffe. According to Rothbard,[95] if they could communicate with us, promise to respect our personal and property rights, and adhere to such undertakings, then and only then would we be obligated to treat them as we do each other (well, better, hopefully).

INTERNATIONAL LAW AND THE SPACE ENVIRONMENT

Despite a relatively large amount of academic writing *about* space law, relatively little space law actually exists for the obvious reason that there has so far been minimal need for it. Still, several international agreements comprise that space law which does exist—and space environmentalists may attempt to use it as a weapon against development. We argue, however, that the international agreements to which the United States is a party do not demand a restrictive environmental regime for outer space and celestial bodies.

The Test Ban Treaty

The Limited Test Ban Treaty of 1963 ("Test Ban Treaty"), signed by the United States, the Soviet Union, and Great Britain, forbids any "nuclear explosion . . . in the atmosphere; beyond its limits, including outer space; or underwater, including territorial waters or high seas," as well as in "any other environment if such explosion causes radioactive debris to be present outside the territorial limits of the State under whose jurisdiction or control such explosion is conducted."[96] Reynolds and Merges describe the treaty as primarily environmental in intent, seeking to prevent "global nuclear contamination," though it had military effects.[97]

We see nothing objectionable in this treaty's ban on the use of nuclear weapons.[98] If states or anyone else want(s) to bind themselves not to conduct certain activities, that seems appropriate. Certainly an inability to detonate nuclear weapons in space will not hamper development there, and should even encourage it to the extent that it reduces the threat of nuclear annihilation of those of us on earth face and eliminates the threat of an electromagnetic pulse that could affect electronic equipment[99] and thereby violate property rights.[100]

The Test Ban Treaty also implicitly prohibits nuclear fission as a means of space propulsion.[101] Again, to the extent that states wish to bind themselves not to do certain things in space, that seems appropriate, just as any private party can agree to refrain from particular actions. The terms of the Test Ban Treaty require the signatories to "prohibit" nuclear explosions "at any place under [their] jurisdiction or control";[102] this would not seem to bind private parties beyond the confines of the Earth. Reynolds and Merges also suggest that if non-signatory nations allow nuclear propulsion, the signatories are likely to follow suit in the interest of competition.[103] To the extent that repudiation of this aspect of the Test Ban Treaty would encourage private parties to use nuclear propulsion, and thereby facilitate space activities, we see it as a laudable step.

The Outer Space Treaty

The Multilateral Treaty on Principles Governing the Activities of States in the Exploration and Use of Outer Space, Including the Moon and Other Celestial Bodies ("Outer Space Treaty")[104] is the foremost document in international space law.[105] The United States and other spacefaring countries have signed it. Primarily, the Outer Space Treaty limits the activities of governments in space. Most significantly, it prohibits them from claiming sovereignty over outer space "including the moon and other celestial bodies,"[106] prohibits "nuclear weapons or any other kinds of weapons of mass destruction" in space,[107] and allows the moon and other celestial bodies to be used "exclusively for peaceful purposes."[108] As with the Test Ban Treaty, we find nothing objectionable here. Governments have been responsible for massive pollution, millions of human deaths,[109] and countless invasions of property rights on Earth; to the extent that they agree to keep their hands off and weapons out of space, that appears favorable from both environmental and property rights perspectives.[110]

Some argue that Article IX of the Outer Space Treaty calls for environmental restrictions.[111] It provides, inter alia, that exploration of celestial bodies must be conducted "so as to avoid their harmful contamination and also adverse changes in the environment of the Earth resulting from the introduction of extraterrestrial matter."[112]

Baker concedes that this rule is essentially anthropocentric—that is, "harmful contamination" simply means contamination that hurts other humans.[113] Nothing in Baker's idea necessarily conflicts with the private-property, strict-liability regime we propose; to the contrary, it is fully consonant with it. And to us it appears to be the most logical reading of the treaty—that is, if contamination of the space environment is, as we have argued, inconceivable except from the perspective of property-rights violations, then the Outer Space Treaty creates no protection for the space environment per se.

Schafer argues that Article VIII of the Outer Space Treaty prohibits contamination of the space environment.[114] Article VIII provides: "Ownership of objects launched into outer space, . . . and of their component parts, is not affected by their presence in outer space . . . ."[115] Schafer reads this provision as establishing "a responsibility to deal with all of the incidence of ownership of such an item, including the contamination it may cause the space environment."[116] Although Schafer is concerned with preserving the environment per se, his argument only supports the notion of a respect for property rights in space. Therefore Article VIII of the Outer Space Treaty is congruent with, not contrary to, our proposed property rights regime and does not establish any environmental protection beyond it.

The Liability Convention

Another major agreement is the 1972 Convention on International Liability for Damage Caused by Space Objects ("Liability Convention").[117] It holds nations strictly liable for damage caused by their (or their citizens') space objects to aircraft in flight or objects on Earth, and applies a "fault" standard for damage to other space objects. The Liability Convention, however, does not address protection of the space environment apart from human-made "space objects" and therefore does not call for any sort of space environmentalism.[118]

The Moon Treaty

A treaty that *does* seem to call for space environmentalism is the 1979 Agreement Governing the Activities of States on the Moon and Other

Celestial Bodies ("Moon Treaty").[119] It requires signatories to "take measures to prevent the disruption of the existing balance of [the moon's] environment, whether by introducing adverse changes in that environment, by its harmful contamination through the introduction of extra-environmental matter or otherwise."[120] It also proclaims the moon and its resources "the common heritage of mankind."[121] Its terms also apply to other celestial bodies.[122]

What precisely the Treaty means by "the existing balance" of celestial body's environment is unclear. As for "harmful contamination," the Moon Treaty does not define the term, but again, in light of our discussion above, this phrase could possibly be interpreted to mean that property rights must be respected.

In any event, the Moon Treaty is of virtually no practical consequence, because "no major space power has signed it."[123] Further, observers agree that it is unlikely that the United States will ever join, because it not only would impose environmental restrictions but also would prohibit property rights[124] and essentially impose a space welfare or global wealth redistribution program, under which profits from wealthy nations' space ventures would have to be shared with the governments of poor non-spacefaring countries.[125] Not even the Soviet Union, despite its communist ideology, was willing to agree to that.[126] Therefore, though it receives attention in academic literature, the Moon Treaty will not directly affect space environmental law in the United States.

CONCLUSION

As we have shown, space environmentalism lacks a legal or economic justification, and its only philosophical foundation is a most extreme form of environmentalism to which very few people seriously subscribe. For the good of the human race, and because it is just, the law should continue to allow private parties to use space for whatever human purposes they see fit within the limits of private property rights.

NOTES

1. Others have been more optimistic about the rate of growth in outer-space activities. See, e.g., Amanda L. Moore and Jerry V. Leaphart, "Manipulation and Modification of the Outer Space Environment: International Legal Considerations," *in Proceedings Of The Twenty-Fifth Colloquium On The Law Of Outer Space* 15, 17 (1982). ("Experts assure that lunar surface mining can be accomplished in the near term future defined as no later than 2000 A.D. and perhaps as soon as 1990.")

2. We include this modifier on the ground that some environmentalists favor private property rights and free enterprise. In our view, "free market environmentalism" is not a contradiction in terms. See Anderson and Leal (1991), Block (1998), Block and Whitehead (1999), DiLorenzo (1990A), Rothbard (1982), Stroup (1991), Whitehead and Block (2000, 2002).

3. In contrast, the present authors characterize themselves as free market private property space environmentalists.

4. Nonetheless, a substantially similar scheme *has* been proposed for the earth. See, e.g., Rothbard, *supra* note 2.

5. For an economic critique of the notion of intrinsic value, *see* Buchanan (1969), Buchanan and Trilby (1981), Barnett (1989), Cordato (1989), DiLorenzo (1990B), Halbrooks (2002).

6. See Flournoy, *supra* note 5, at 81–82; John Grim, "Indigenous Traditions and Deep Ecology", *in* Deep Ecology and World Religions: New Essays on Sacred Ground 35, 40–41 (David Landis Barnhill ed., 2001).

7. Mikael Stenmark, *Environmental Ethics and Policy Making* 85 (2002); see also Holmes Rolston, III, "Ethics on the Home Planet," *in An Invitation to Environmental Philosophy* 107, 133 (Anthony Weston ed., 1999) ("Earth does not belong to us; rather we belong to it. . . . Earth is really the relevant survival unit.").

8. The land ethic "enlarges the boundaries of the community to include soils, waters, plants, and animals, or collectively: the land." Gary D. Meryers, Old-Growth Forests, the Owl, and Yew: Environmental Ethics Versus Traditional Dispute Resolution Under the Endangered Species Act and Other Public Lands and Resources Laws, 18 *B.C. Envtl. Aff. L. Rev.* 623, 657 (quoting Aldo Leopold, *A Sand County Almanac* 202–03 [1949]).

9. See George Reisman, "Environmentalism in the Light of Mises and Menger," 5, *Q. J. Austrian Econ.* 3, 11 (Summer 2002).

10. Block, *Environmentalism and Economic Freedom: The Case for Private Property Rights*, *supra* note 2, at 1896 n5.

11. Robert James Bidinotto, "Environmentalism: Freedom's Foe for the 90s," *The Freeman*, November 1990 (quoting M. John Fayhee, "Earth First! and Foremost," *Backpacker*, September 1988, at 21).

12. Stenmark, *supra* note 12, at 85.

13. *Id.* at 89.

14. See, e.g., Mary Anne Warren, "The Rights of the Nonhuman World," in *Environmental Philosophy: A Collection of Readings* 109 (Robert Elliot and Arran Gare eds., 1983) (stating that environmentalists show concern for the "planetary biosystem" and the "biotic community" as a whole).

15. Erazim Kohák, *The Green Halo: A Bird's-Eye View of Ecological Ethics* 129 (2000).

16. Ulrike M. Bohlmann, "Planetary Protection in Public International Law," in *Proceedings of the Forty-Sixth Colloquium on the Law of Outer Space* 18, 27 (2003).

17. See Ryder W. Miller, "Astroenvironmentalism: The Case for Space Exploration as an Environmental Issue," 15 *Electronic Green J.* (2001), available at http://egj.lib.uidaho.edu/egj15/miller1.html.

18. Howard A. Baker, "Protection of the Outer Space Environment: History and Analysis of Article IX of the Outer Space Treaty," in 12 *Annals of Air & Space L.* 143,166 (Nicolas Mateesco Matte, ed., 1987).

19. Bernard K. Schafer, "Solid, Hazardous, and Radioactive Wastes in Outer Space: Present Controls and Suggested Changes," in *Law, Values, and the Environment: A Reader and Selective Bibliography* 395, 399 (Robert N. Wells, Jr. ed., 1996) (emphasis added). One wonders whether the "future generations" of which he speaks are intended to be human beings. If so, this represents an inconsistency in his world view, which, presumably, denigrates human beings of whatever generation, now *or* in the future. This author seems to think humans (now and in the future) can "enjoy" outer space—apparently by looking at it and knowing it is there—as long as they leave it in its pristine state. But, people cannot own views. If they did, there would be an over-determination of property rights: millions of people would own Disney World and the Grand Canyon, and it would no longer be clear as to who had rights to alter these amenities. See Walter Block, "Homesteading, AD Coelum, Owning Views and Forestalling," (2008).

20. April Greene Apking, Note & Comment, "The Rush to Develop Space: The Role of Spacefaring Nations in Forging Environmental Standards for the Use of Celestial Bodies for Governmental and Private Interests," 16 *Colo. J. Int'l Envtl. L. & Pol'y* 429, 433 (2005).

21. Even NASA is calling for more private initiatives in space travel and exploration. See Lucy Sherriff, "Private Enterprise Needed in Space: NASA," *The Register,* November 17, 2005, available at http://www.theregister.co.uk/2005/11/17/nasa_private_investment/; see also George Knapp, "The Ultimate Public-Private Partnership, Las Vegas Mercury," July 8, 2004, available at http://www.lasvegasmercury.com/2004/MERC-Jul-08-Thu-2004/24250261.html. On the other hand, nothing in this chapter should be interpreted as our support for any such initiative. In our view, the market should be left alone to determine just how much investment should go to this area. Certainly, no government subsidies would be justified. For a general critique of NASA as a socialist enterprise and support of private entrepreneurial decision making in this arena, see, for example, William L. Anderson, "The Trouble with NASA," *The Free Market*, April 2003, available at http://www.mises.org/freemarket_detail.asp?control=434. See also Robert Murphy, "A Free Market in Space," *The Free Market*, January 2005, available at http://www.mises.org/freemarket_detail.asp?control=525.

22. Miller, *supra* note 22.

23. This suggests a legal system in which rocks, dirt, and other inanimate objects somehow have standing to sue those who have disrupted them. Presumably environmentalists would bring these claims on the rocks' behalf, as they did for trees in *Sierra Club v. Morton*, 405 U.S. 727 (1972). On that case, and some economic problems of standing to sue for "existence value," see Donald J. Boudreaux and Roger E. Meiners, "Existence Value and Other of Life's Ills," in *Who Owns the Environment?* 153, 177–79 (Peter J. Hill and Roger E. Meiners eds., 1998).

24. Paul F. Uhlir and William P. Bishop, "Wilderness and Space," in *Beyond Spaceship Earth: Environmental Ethics and the Solar System* 183, 203–04 (Eugene C. Hargrove ed., 1986) (citing Harry H. Almond, Jr., "A Draft Convention for

Protecting the Environment of Outer Space," in *Proceedings of the Twenty-Third Colloquium on the Law of Outer Space* 101–02 (1980)) (emphasis added).

25. *Id.* at 205.

26. Glenn H. Reynolds & Robert P. Merges, *Outer Space: Problems of Law and Policy* 176 (1997).

27. *Id.*

28. Lawrence D. Roberts, "Ensuring the Best of All Possible Worlds: Environmental Regulation of the Solar System," 6 *N.Y.U. Envtl. L.J.* 126, 153, 158–59 (1997).

29. See, e.g., Steven Freeland, "Up, Up and . . . Back: The Emergence of Space Tourism and its Impact on the International Law of Outer Space," 6 *Chi. J. Int'l L.* 1, 20 (2005) (declaring laws against "littering" by space tourists "imperative" to avoid "additional disruption to the space environment"); Kelly M. Zullo, Note, "The Need to Clarify the Status of Property Rights in International Space Law," 90 *Geo. L.J.* 2413, 2442 (2002). ("Renewable licenses [for space ventures] should be granted liberally unless . . . the proposed activity would cause an unacceptable degree of harm to the Earth or outer space environment. An unacceptable degree of harm may include ventures that would leave excessive debris, produce harmful radioactive waste, or some other demonstrable damage.") For an alternative view on littering, see *Walter Block, Defending the Undefendable*, 210–16 (Fox & Wilkes 1991) (1976).

30. See *supra* note 9.

31. See Ehrlich, *supra* note 7; Ehrlich & Ehrlich, *supra* note 7.

32. Mark A. Garlick, *The Story of the Solar System* 52 (2002) (emphasis added).

33. *Id.* at 58.

34. *Id.*

35. *Id.* at 58, 60.

36. *Id.* at 60. It may seem strange that Venus is further from the sun than Mercury, and, yet, hotter. This is due to its thick atmosphere of carbon dioxide which creates a "greenhouse effect."

37. See *id.* at 66.

38. *Id.* at 72.

39. *Id.* at 74.

40. *Id.* at 72.

41. See *id.* at 86–106.

42. *Id.* at 112.

43. See Kenneth Chang and Dennis Overbye, "Planet or Not, Pluto Now Has Far-Out Rival," *N.Y. Times*, July 30, 2005, at A1.

44. Garlick, *supra* note 37, at 97.

45. Schafer, *supra* note 24, at 427 n.19 (citing Michael Freeman, *Space Traveller's Handbook: Every Man's Comprehensive Manual to Space Flight* 154 (1979)).

46. Reynolds & Merges, *supra* note 31, at 13.

47. Garlick, *supra* note 37, at 127–42. Lest we depress our readers too severely, we note that author Robert Ringer suggests using this "ice ball" scenario as an

opportunity to recognize that one's day-to-day problems are relatively insignificant. See Robert J. Ringer, *Winning Through Intimidation* 40–43 (1974).

48. Another antidote to depression: if we do not blow each other up before that, it is at least possible that long before this time man will have learned the techniques necessary to bring us to not only other planets but other solar systems, more welcoming ones.

49. On the enmity between environmentalists and carbon dioxide on Earth, see, for example, Gore, *supra* note 7, at 22.

50. Nat'l Research Council, Safe on Mars: Precursor Measurements Necessary to Support Human Operations on the Martian Surface 28 (2002), available at http://newton.nap.edu/books/0309084261/html/28.html.

51. Benjamin Wallace-Wells, "Mars or Bust," *Rolling Stone*, February 23, 2006, at 45, 50.

52. Roberts, *supra* note 33, at 154.

53. In fairness to Roberts, he recognizes this point at least to a limited extent. See *id.* at 149 ("With regard to the inhospitable resources of the solar system, any environmental regime should begin with the management of property rights."); *Id.* at 151 ("Enforcement, in most cases, is a matter of individual property owners upholding their personal interests.").

54. See Rothbard, *supra* note 2, at 88.

55. Richard A. Posner, *Economic Analysis of Law* § 13.5 (5th ed. 1998).

56. For example, if courts had continued to allow individuals to bring lawsuits for air pollution from the nineteenth century through the present, an environmental forensics industry would almost certainly be capable of determining sources of harm from air pollution. See Morton Horwitz, *The Transformation of American Law: 1780-1860* (1977); Walter Block, "Private Property Rights, Economic Freedom, and Professor Coase: A Critique of Friedman, McCloskey, Medema, and Zorn," 26 *Harv. J.L. & Pub. Pol'y* 923, 929–30 (2003).

57. Roberts, *supra* note 33, at 154.

58. A relatively low number of parties makes bargaining more likely (though perhaps not certain if transaction costs are high). See Posner, *supra* note 61, at § 3.8; Ronald H. Coase, "The Problem of Social Cost," 3 *J.L. & Econ.* 1 (1960).

59. See Rothbard, *supra* note 2, at 77.

60. Such an act would be characterized as "coming to the nuisance."

61. See Block, *Environmentalism and Economic Freedom: The Case for Private Property Rights*, *supra* note 2, at 1890; see also Horwitz, *supra* note 62.

62. See, e.g., Petr Beckmann, *The Health Hazards of NOT Going Nuclear* 122 (1977). ("Every 1,000 MW of nuclear power that replaces coal-fired power saves between 20 and 100 lives a year.")

63. See *id.* at 102 (noting also that while it is correct to say that nuclear waste remains hazardous for thousands of years, the problem is not as threatening as it might seem, nor as threatening to human health in the short or long term as the burning of fossil fuels).

64. See Ty S. Twibell, Note, "Space Law: Legal Restraints on Commercialization and Development of Outer Space," 65 *UMKC L. Rev.* 589, 631 (1997).

65. See Beckmann, *supra* note 68, at 99–111 ("If all of the US power capacity were nuclear, the total amount of wastes per person per year would amount to one aspirin tablet. . . . If the entire US electrical capacity were nuclear and ran at the present [1977] rate for 350 years, the volume of wastes would amount to a cube 200 feet on a side.").

66. For example, NASA lost two space shuttles and 14 astronauts in just 114 flights. Tariq Malik, "NASA's New Moon Plans: 'Apollo on Steroids,'" *Space.com,* September 19, 2005, http://www.space.com/news/050919_nasa_moon.html. In 1999 alone, it also lost two Mars spacecraft and a $246 million infrared telescope which failed shortly after launching. "Nasa's Disastrous Year," *BBC News*, March 22, 2000, http://news.bbc.co.uk/1/hi/sci/tech/686674.stm.

67. See Kenneth Chang, "Not Science Fiction: An Elevator to Space," *N.Y. Times,* September 23, 2003, at F1.

68. See Roberts, *supra* note 33, at 148.

69. Reynolds & Merges, *supra* note 31, at 176. See also Robert P. Merges and Glenn H. Reynolds, "Space Resources, Common Property, and the Collective Action Problem," 6 *N.Y.U. Envtl. L. J.* 107, 124–25 (1997). Incidentally, Reynolds and Merges also call for "development preserves"—territory set aside to give to poor non-spacefaring nations later. Reynolds & Merges, *supra* note 31, at 176.

70. Reynolds & Merges, *supra* note 28, at 176.

71. See, e.g., Miller, *supra* note 18; Schafer, *supra* note 24.

72. These groups include, for example, the Nature Conservancy and the Audubon Society. See John Brätland, "Externalities, Conflict, and Offshore Lands," 8 *Indep. R.* 527, 540 (2004).

73. See *id.*

74. On the ability of polluters to "capture" centralized agencies ostensibly intended to regulate them, see, for example, Andrew P. Morriss, Bruce Yandle and Roger E. Meiners, "The Failure of EPA's Water Quality Reforms: From Environment-Enhancing Competition to Uniformity and Polluter Profits," 20 *UCLA J. Envtl. L. & Pol'y* 25, 26 (2001–2002).

75. See, *e.g.*, Block, *Environmentalism and Economic Freedom: The Case for Private Property Rights*, *supra* note 2, at 1889; Fred L. Smith, Jr., "Sustainable Development–A Free-Market Perspective," 21 *B.C. Envtl. Aff. L. Rev.* 297, 305–06 (1994).

76. Glenn Harlan Reynolds, "Space Law in the 21st Century: Some Thoughts in Response to the Bush Administration's Space Initiative," 69 *J. Air L. & Com.* 413, 422 (2004).

77. *Id.* at 419.

78. *Id.* at 420 (citing Martyn J. Fogg, *Terraforming: Engineering Planetary Environments* [1995]).

79. *Id.* at 420.

80. *Id.*

81. *Id.* (quoting Robert Zubrin, *Entering Space: Creating a Spacefaring Civilization 37* [1999]).

82. *Id.*

83. A battle between "reds" who want to preserve a pristine Mars and "blues" who want to terraform occurs in an acclaimed series of science-fiction novels. See Kim Stanley Robinson, *Red Mars* (1992); Kim Stanley Robinson, *Blue Mars* (1993); Kim Stanley Robinson, *Green Mars* (1996).

84. See Reynolds, *supra* note 82, at 421 ("The character of these objections is likely to reveal much about the environmental movement, or at least about those making them."); see also, e.g., Ryder W. Miller, 'The Case Against Terraforming Mars," *Space Pol'y Dig.*, May 22, 2003, available at http://www.gyre.org/news/article/3273.

85. See John Locke, "An Essay Concerning the True Origin, Extent and End of Civil Government," *in Social Contract* 17–18 (E. Barker, ed., 1948); see also Hans-Hermann Hoppe, *The Economics and Ethics of Private Property: Studies in Political Economy and Philosophy* (1993); Murray N. Rothbard, *The Ethics of Liberty* 32 (New York University Press, 1998) (1982); Murray N. Rothbard, *For a New Liberty* 34–35 (Ludwig von Mises Institute, 2002) (1973); Walter Block, "Earning Happiness Through Homesteading Unowned Land: A Comment on 'Buying Misery with Federal Land' by Richard Stroup," 15 *J. Soc. Pol. & Econ. Stud.*, 237 (1990).

86. For an analogous case regarding intellectual property, see N. Stephan Kinsella, "*Against Intellectual Property*," 15 *J. Libertarian Stud.* 1 (2001) (explaining that the first creator of an idea has an advantage in reaping the economic benefits from the idea even if he cannot own it).

87. For a rejection of the notion of a right to weather of a particular type in the context of global warming on Earth, see Reisman, *supra* note 14, at 14.

88. See *supra* note 9.

89. See Dennis Overbye, "When It's Not Enough to Say 'Take Me to Your Leader,'" N.Y. Times, March 5, 2002, at F1.

90. Roberts, *supra* note 33, at 157–60. For more on the precautionary principle, see Steven A. Mirmina and David J. Den Herder, "Nuclear Power Sources and Future Space Exploration," 6 *Chi. J. Int'l L.* 149, 164 (2005).

91. See Bruce Jakosky, *The Search for Life on Other Planets* 2 (1998).

92. *Id.* at 4.

93. The most earthlike planet known is 21,000 light-years away and has a surface temperature of negative 370 degrees Fahrenheit. Dennis Overbye, "Search Finds Far-Off Planet Akin to Earth," *N.Y. Times*, January 26, 2006, at A21.

94. This has happened on Earth. For example, the Endangered Species Act encourages people to kill and hide any endangered species on their property (in the vernacular, "shoot, shovel, and shut up") so they do not lose their property rights. See Andrew P. Morriss and Richard Stroup, "Quartering Species: The 'Living Constitution,' the Third Amendment, and the Endangered Species Act," 30 *Envtl. L.* 769, 795 (2000).

95. Rothbard, *For a New Liberty*, *supra* note 91, at 156; see also Tibor Machan, *Putting Humans First: Why We Are Nature's Favorite* (2004).

96. Treaty Banning Nuclear Weapon Tests in the Atmosphere, in Outer Space and Under Water, U.S.-U.K.-U.S.S.R., art. I, August 5, 1963, 14 U.S.T. 1313 [hereinafter Test Ban Treaty].

97. Reynolds & Merges, *supra* note 31, at 54.

98. For a libertarian analysis of weapons in space and on the heavenly bodies in a different than earthly context, *see* Walter Block and Matthew Block, "Toward a Universal Libertarian Theory of Gun (Weapon) Control," 3 *Ethics, Place & Env't* 289 (2000).

99. See Reynolds & Merges, *supra* note 31, at 59.

100. Former U.S. Secretary of State Dean Rusk noted, "I can think of no other way to more massively increase the pollution of outer space than to allow the arms race to move out there." Dean Rusk, "Star Wars: The Nuclear/Military Uses of Space," *in Beyond Spaceship Earth: Environmental Ethics and the Solar System* 315, 318 (Eugene C. Hargrove ed., 1986).

101. Reynolds & Merges, *supra* note 31, at 61.

102. Test Ban Treaty, *supra* note 102, art. I.

103. Reynolds & Merges, *supra* note 31, at 61.

104. Treaty on Principles Governing the Activities of States in the Exploration and Use of Outer Space, Including the Moon and Other Celestial Bodies, U.S.-U.K.-U.S.S.R., January 27, 1967, 18 U.S.T. 2410 [hereinafter Outer Space Treaty].

105. See Steven A. Mirmina and David J. Den Herder, "Nuclear Power Sources and Future Space Exploration," 6 *Chi. J. Int'l L.* 149, 158 (2005) (calling the Outer Space Treaty "the cornerstone of international space law"); Reynolds & Merges, *supra* note 31, at 62.

106. Outer Space Treaty, *supra* note 110, art. II.

107. *Id.* art. IV.

108. *Id.*

109. See R.J. Rummel, *Death by Government* (1994). New Brunswick, NJ: Transaction.

110. Some have argued that the Outer Space Treaty effectively abolishes both government and private property in space. Reynolds, however, argues that a consensus exists to the contrary—that is, that property rights are allowed. Glenn H. Reynolds, "International Space Law: Into the Twenty-first Century," 25 *Vand. J. Transnat'l L.* 225, 230 (1992). But the idea that the Outer Space Treaty proscribes private property persists among some commentators, and the status of private property rights in space therefore remains uncertain. See, e.g., Heidi Keefe, Essay, "Making the Final Frontier Feasible: A Critical Look at the Current Body of Outer Space Law," 11 *Santa Clara Computer & High Tech. L.J.* 345, 350–60 (1995); Leo B. Malagar and Marlo Apalisok Magdoza-Malagar, "International Law of Outer Space and the Protection of Intellectual Property Rights," 17 *B.U. Int'l L.J.* 311, 345 (1999); Uhlir and Bishop, *supra* note 29, at 196 (noting that this prohibition will preserve space "wilderness"). This debate is beyond the scope of this article, but we note that a considerable consensus does exist that, even if the Outer Space Treaty does not allow for private property rights, it *should* allow for them so entrepreneurs will have an incentive to go there. See, e.g., Kurt Anderson Baca, "Property Rights in Outer Space," 58 *J. Air L. & Com.* 1041, 1083–85 (1993); Julie A. Jiru, Comment, "Star Wars and Space Malls: When the Paint Chips Off a Treaty's Golden Handcuffs," 42 *S. Tex. L. Rev.* 155, 169–73 (2000); Twibell, *supra* note 70, at 613–19; Keefe, at 350–60; Wayne N. White, Jr.,

"Real Property Rights in Outer Space," in *Proceedings of the Fortieth Colloquium on the Law of Outer Space* 370 (1998); Zullo, *supra* note 34, at 2438–44.

111. See, e.g., Molly K. Macauley, "Flying in the Face of Uncertainty: Human Risk in Space Activities," 6 *Chi. J. Int'l L.* 131, 144 (2005); Uhlir and Bishop, *supra* note 29, at 196–97.

112. Outer Space Treaty, *supra* note 110, art. IX.

113. See Baker, *supra* note 23, at 163 ("[I]t was never intended that the protection offered by the avoidance of harmful contamination principle would extend to the environments of the Moon and other celestial bodies *per se*.").

114. Schafer, *supra* note 24, at 404.

115. Outer Space Treaty, *supra* note 110, art. VIII.

116. Schafer, *supra* note 24, at 404.

117. Convention on International Liability for Damage Caused by Space Objects, U.S.-U.K.-U.S.S.R., March 29, 1972, 24 U.S.T. 2389 [hereinafter Liability Convention].

118. This is in sharp contrast to the Convention on International Trade in Endangered Species of Wild Fauna and Flora ("CITES"), March 3, 1973, 27 U.S.T. 1087, the aqueous analog to the Liability Convention. CITES specifically rules out and condemns private property rights as a means of protecting endangered species. See Jonathan Adler, "Do Conservation Conventions Conserve?," in *Sustainable Development: Promoting Progress or Perpetuating Poverty?* (Julian Morris ed., 2002).

119. Agreement Governing the Activities of States on the Moon and Other Celestial Bodies, G.A. Res. 34/68, U.N. Doc. A/RES/34/68 (December 5, 1979) [hereinafter Moon Treaty].

120. *Id.* art. 7.

121. *Id.* art. 11. For an analysis of this concept, see Kemal Baslar, *The Concept of the Common Heritage of Mankind in International Law* (1998); for applications of this concept to the earth's oceans, see Jan van Ettinger et al., "Ocean Governance and the Global Picture," in *Ocean Governance: Sustainable Development of the Seas* (Peter Bautista Payoyo ed., 1994); for a critique on the ground that it interferes with private property rights, and opens the door to the tragedy of the commons, see Roy Whitehead, Jr., Catherine Gould and Walter Block, "The Value of Private Water Rights: From a Legal and Economic Perspective," 9 *Alb. L. Envtl. Outlook J.* 313 (2004).

122. Moon Treaty, *supra* note 125, art. 1.

123. Rosanna Sattler, "Transporting a Legal System for Property Rights: From the Earth to the Stars," 6 *Chi. J. Int'l L.* 23, 30 (2005); see also Reynolds & Merges, supra note 31, at 116. ("Absent adoption by the major space powers, the Moon Treaty is unlikely to play a major role in the future.")

124. See Sattler, *supra* note 129, at 30; Twibell, *supra* note 70, at 598.

125. See Kevin V. Cook, Note, "The Discovery of Lunar Water: An Opportunity to Develop a Workable Moon Treaty," 11 *Geo. Int'l Envtl. L. Rev.* 647, 664–70 (1999).

126. Reynolds and Merges, *supra* note 31, at 116.

Chapter 8

Land for Parks

PUBLIC PARKS: THE NEW YORK CITY CASE

Everyone knows that the system of public parks in New York City is a mess. Our city parks have become havens for muggers and junkies, unsafe for honest citizens except perhaps in broad daylight on summer weekends. With the exception of Central Park, virtually all the city parks are spread around the periphery of Manhattan, where they are all but impossible to reach. Tourists on the Circle Line boat cruise are likely to conclude that virtually all of Manhattan is composed of parks; but we occupants of the barren inland know better, unfortunately. As for the pitifully few inland "vest pocket parks," they are in danger of becoming obsolete.

In Tudor City, a group of elderly people have formed the Save Our Parks Committee, to stop Tudor City's owner, the Helmsley Spear real estate firm, from erecting a 52-story residential and office building on the site of two small parks on East 42nd Street near the U.N. Predictably, the politicians of the area have rallied to the cause of saving the park, and have criticized the greedy profit-seeking landlord. The problem with the facile answer of supporting the park however is that on the one hand, we also have a housing shortage and perhaps the park should be replaced by housing. On the other hand, it may indeed be true that the park shortage is even greater than the housing shortage, and then we should not only save these small parks but should replace some housing with parks. The real problem of the park system in New York City is not whether we should save these two small parks, important as that may be. It is that we lack a mechanism by which these and all similar problems can be solved automatically.

When the problem of how to allocate our scarce resources among competing ends arises in other areas, we have an answer that has served us well.

It has served us so well, in fact, that it can only be folly to ignore its application to the present problem. I speak, of course, of the price system. If there were ever to arise a problem of allocating scare resources between shoes and wheat of the sort that has arisen between parks and housing, where a "shoes faction" opposed a "wheat fraction," the price system would solve this problem in short order. (Indeed, the very ludicrousness of the example strongly suggests that the price system does not even allow such problems to arise in the first place.) For the good in relatively short supply would rise in price as unsatisfied demanders bid for it. If wheat were more scarce, the price of wheat would rise relative to shoes. Then wheat producers would be able to bid land and labor and other scarce factors of production away from the shoe producers.

How would the price system work in the case of parks? The city government would denationalize the park industry. It would sell off all its park lands to private entrepreneurs who would have the option of maintaining the land in the form of a park or converting it to some other use. These entrepreneurs would charge prices for the use of parks just as is done in the form of single-entry admissions, season tickets, or any other plan amenable to both park owner and customer.

The advantages of the plan would be immediate. No longer would the park-using public have to fear for its very life. What would solve the problem of crime in the parks would not be simply more police or better lighting or any of the other specific measures often proposed by the bureaucrats in charge of parks. What will solve the problem is rather a system which will automatically reward those entrepreneurs who are able to rid the parks of crime by whatever methods they adopt, and a system which will automatically penalize those entrepreneurs who are unable to rid the parks of crime. The profit and loss system, or the price system, will give rise to a whole host of park owners, each free to use his own methods. Those who succeed will be rewarded by the patronage of the customers, will prosper, and will be able to spread their enlightened methods to other parks. Those who fail will lose customer support, will lose money and go bankrupt, and will no longer be in position to mismanage parks. It is in this way that the price system will improve the safety of parks.

The same analysis holds true for other facets of park operation: the location of the park in the first place, the other services provided by park managers, and so on: those entrepreneurs who please the customers will prosper, and those who do not will fail. It is in this way, and this way alone, that we are likely to have the parks relocated to where they are more accessible, that we are likely to see the parks open at night, and so on.

The radical thing about this proposal is not the use of the price system itself, which is used in all aspects of our economic life, but rather the application of it to an area in which it had never been used before. Unfortunately, we are such slaves to the past that it is extremely difficult to imagine alternative ways of doing things, where the alternatives would replace institutions that have been with us for many years. Objection to new ways arise which would never occur to us but for their newness. For example, if the government had been in the business of providing shoes and wheat in much the same way that it is now in the business of providing parks, fire protection, postal service, and so on, and if someone were to come forward with a proposal to turn the production of shoes and wheat over to private enterprise, all sorts of objections would probably arise which from our present vantage point of experience would appear to be frivolous: "How would the farmers bring the wheat to market?" "Who would sew the sales and shoe tops together?" "How would the merchants be able to charge for the shoes?" "Who would bake the bread?" "How would we decide on the proportion of wheat allocated to bread and to cake?" and especially, "But the poor would be forced to go without shoes!" From the vantage point of experience, we know all these objections about shoes and wheat to be baseless. But they are very potent indeed with regard to the de-nationalization of parks, an action where we have no experience to guide us.

Let us consider in some detail the claim that the poor would be forced to go without the use of parks, since we have dealt with the other objections with regard to parks, at least briefly. It is true that poor people and poor children (along with everyone else) will be forced to pay for the park services they consume under a system of private ownership of the parks, just as they are now forced to pay for the shoes and wheat they consume. But it by no means follows that the poor will not be able to enjoy the use of parks, or even that they will be forced to curtail their use of parks! If anything, the de-nationalization of our parks will probably mean more park use by the poor, not less.

For one thing, the poor, along with everyone else, will be able to make better use of private parks because they will tend to be run more efficiently under the price system. No longer will the parks be out of service because of uncontrolled crime, improper lighting, or defective equipment. Government bureaucratic park managers receive their salary regardless of consumer satisfaction; entrepreneurs do not. But more importantly, we must realize that the effect of the present system of nationalized parks is to divert recreational funds away from poor people and toward rich people. Therefore, denationalizing the park industry will have the effect of increasing, not decreasing, the amount of park services that go to the poor!

Although it is true that the New York City government does spend some thousands of dollars on building and maintaining asphalt playgrounds and swimming pools used mainly by the poor, it severely taxes recreations mainly attended by the poor, such as professional baseball, football, and basketball games, bowling alleys and pool halls, and so on. The city government then takes these and other tax funds and subsidizes recreational activities mainly patronized by the rich to the tune of literally billions of dollars. The priceless art treasures in our museums and art galleries, the millions of dollars that go into our libraries, ballet and opera companies, symphony orchestras, Shakespeare Festivals, and so on, all represent recreational money taken away from the poor and given to the rich. Can it then be seriously maintained that stopping this process by getting the government out of the recreation business will actually be detrimental to the poor? Hardly.

No. The answer to the physical and spiritual decay of our nationalized recreational industry can only be a separation of recreation and state akin to the separation of church and state mandated by the constitution. And such a separation can only help the poor, the greatest victims of state depredations.

Section V

ENVIRONMENTAL ECONOMICS

Chapter 9

The Economics of the Arctic National Wildlife Refuge

INTRODUCTION: ARCTIC NATIONAL WILDLIFE REFUGE: A CASUALTY OF LEASING LAW OR CANDIDATE FOR PRIVATIZATION?

The government has been withholding opportunities for private owners to allocate resources efficiently and effectively by enacting legislation that allows continuous bureaucratic control over valuable resources. The Arctic National Wildlife Refuge (ANWR) in Alaska is a poignant example of how the government has restricted the market, and indirectly consumers as a whole,[1] in deciding how natural resources are to be used. Individual buyers and sellers in the marketplace are not given the chance to decide resource allocation in this case. Instead, the ownership remains in the hands of bureaucrats and the choice of how to use the resources is left to the skewed political process of special interest groups and special interest political agendas.

Arctic National Wildlife Range (http://pubs.usgs.gov/fs/fs-0028-01/fs-0028-01.htm) was created in the 1960s and designated as government-owned property "for the purpose of preserving unique wildlife, wilderness and recreational values" (http://www.r7.fws.gov/nwr/arctic/purposes.html). The Alaska National Interests Conservation Act (ANILCA) of 1980 followed, which enlarged government ownership to 100 million acres. Nineteen million of those acres comprise the ANWR. We focus on this area of land, and specifically the 1.5 million-acre coastal plain called Area 1002, for two reasons —wildlife and oil.

Drilling for oil in ANWR is supported by those who feel strongly about decreasing domestic dependence on foreign energy sources. With the war in Iraq, and the overall political volatility in the Middle East, the major supplier of the world's oil, it is important for the United States to consider expanding domestic production.[2] In addition, the infrastructure needed to transport the oil to U.S. markets is already in place in the nearby Prudhoe Bay. This area has already been open to leased oil drilling, and much has been learned about how to minimize the risks associated with recovering oil in the Arctic (Lee, 2001, p. 217).

On the other hand, ANWR is considered by anti-market environmentalists to be as valuable pristine, untouched wilderness. Both the scenic landscapes and the wildlife within the area have "existence value" of their own (Anderson, 2002, p. 1), it is claimed. Drilling for oil is not worth disrupting the ecosystem contained within ANWR, especially because the amount of oil is uncertain, and any mistake made in the process is seen by socialist environmentalists as unacceptable. These left-wing supporters think that forcing conservation measures on individuals is the only way to accomplish their goals.

ENVIRONMENTAL SOCIALISM

"Existence value" is defined as "the value that individuals may attach to the mere knowledge of the existence of something, as opposed to having direct use of that thing" (http://economics.about.com/library/glossary/bldef-existence-value.htm).

"Existence value" compounds the difficulties of calculation pointed out by the Austrians in the socialist calculation debate (Anderson, 2002, p. 2; Ebeling, 1993; Hayek, 1948; Herbener, 1996; Hoff, 1981; Horwitz, 1996; Keizer, 1987, 1997; Kirzner, 1988; Klein, 1996; Lavoie, 1981, 1985; Lewin, 1998; Pasour, 1983; Rothbard, 1971, 1976; Salerno, 1990, 1995; Steele, 1981, 1992). For here, it is not a "mere" matter of determining which is more valuable, say, steel in cars or in tractors; or, to keep closer to our present concerns, caribou, or oil pipelines. With full private property rights, markets solve such problems based on private property and profits and losses every day. But "existence value," by its very nature, is not amenable, even in principle, to any such solution. For in all these other cases, the interests of the owners of one good (steel cars, caribou) may be set against those of the other (steel tractors, oil pipelines) and the presumption is that the one set of owners able to outbid the other for the scarce resource will determine maximum value usage. However, in the case of "existence value," there are by definition no two sets of owners, and scarce resources to be allocated between them. Rather, according to the advocates of this curious doctrine, anyone and everyone, without exception, property owners and non-property owners, are somehow, magically, allowed to attain for themselves the right to determine land usage.

A rational allocation of resources is based upon people bidding for things such as steel tractors and oil pipelines, and their owners, attempting to maximize profit, selling or renting them to the highest bidders, ceteris paribus. In that way, we may expect that the market will always and everywhere tend in the direction of allocating resources in accordance with the wishes of all system participants. But it is quite another matter when one of the resources to be allocated is not *owned* by anyone, where all people may in effect claim property rights, for example, assert the right to make decisions as to the "existence value" of land they had not purchased nor homesteaded. Then, all bets are off and the economy is if at sea without a rudder. For one thing, different people may have opposite "existence values" concerning the same resource. To watermelon[3] environmentalists, the ANWR may be a pristine wilderness, deserving of protection at pretty much all and any costs. To others, this territory might seem a vast waste, fit only for paving over and thus bringing into economic civilization. Who is to choose, given that there are no owners?[4] For another something akin to this sort of voting, economic democracy, as it were, was already tried, in the USSR and Eastern Europe. There, it was judged a gigantic failure. What is the case, then, for employing such a model in the "land of the free, home of the brave?" This is not to say, of course, that political democracy was prevalent in these polities. Nothing could be further from the truth. However, democratic rule over industry is *economically* indistinguishable from the Soviet central planning. The only difference is the size of the central planning board. In one case, at least theoretically, it is the entire country; in the other, it is limited to a small oligarchy. But as far as prices, profits and losses, private property are concerned, the be all and end all of rational economics, they are all absent, equally absent, in both cases.[5]

The argument for conservation or drilling can only be rationally decided in the context of private ownership. The government currently owns the land that is being disputed. However, as we have seen, this is a recipe for, at best, centralized socialist planning. As a result, the best use of ANWR simply cannot be determined because accurate economic calculation can only occur if the property is privately owned. Without a profit and loss system, and private property, there is no sure way to identify whether the benefits of drilling for oil outweigh that of the wildlife and landscape. Assertions of "existence value," however, can be used to justify (or nullify) anything without providing a mechanism by which the claims can be substantiated.[6]

The best course of action regarding oil vs. caribou is difficult to determine because it involves subjective values. Some environmentalists subjectively value the scenic landscapes and diverse wildlife. For such people, drilling for oil may be regarded as an ugly distortion to the Arctic environment. Another person looking at the same picture may be impressed with the cooperation between human ingenuity and nature's beauty. Yet others may not care at all about the wildlife, particularly if their job depends on oil exploration and development. Subjective values are easily reflected within a free market system of prices, and allocation of these valuable resources is most efficiently achieved by Adam Smith's "invisible hand."

Mises (1933, 1969) argued that the pricing systems in socialist economies were necessarily deficient because if the government owns the means of production, as our government does in the case of ANWR, then correct pricing of these goods cannot be generated. One of the reasons for this is that goods in the hands of government are merely internal transfers, and not subject to the pricing methods of the market (http://cepa.newschool.edu/het/paretian/social.htm). Anderson and Leal (2001, p. 77) emphasize the effects of this internal transfer concept with regard to resource allocation within ANWR:

> Recognizing that politics will determine the distribution of values from the energy and environmental resources, both sides will invest time and money lobbying government. But lobbying does not create wealth; it only redistributes it. With many energy resources in the hands of state and federal bureaucrats, the transfer game has become extremely important for both oil companies and environmental groups, which realize that their wealth will be determined by bureaucratic decisions.

Hayek (1945) maintained that the central planning system required calculations that were essentially impossible because they required information that is not attainable by the "public servants." Without an accurately functioning price system, it is simply impossible for any central authority to gather the information which exists only in the minds of millions of market participants. In addition, "the economic incentives provided in a market economy could not be reproduced in a collective system" (http://cepa.newschool.edu/het/paretian/social.htm).

To summarize to this point: "existence value" is a will-o-the-wisp. It is not conducive to a bargaining process, to markets, which are the bedrock of our economy. Suppose, we posit it to exist, arguendo. That still does not address the issue of *whose* "existence value." Markets and private property rights allow for a process of bidding, thereby providing the only means we have to peacefully and non-arbitrarily resolve such conflicts.

GOVERNMENT LEASING VS. OWNERSHIP

One must acknowledge the fact that allocation decisions made within both the government and the market are invariably based on asymmetric information. In 1987 the U.S. Department of Interior conducted the ANWR, Alaska,

Coastal Plain Resource Assessment. This 227-page study is just one of hundreds of similar documents that compiles available information by these "public servants" in order to assist the government bureaucrats in making allocation decisions. Hodel (1987) prefaces this document as follows:

> At that time [during the creation of ANILCA, 1980], the Congress specifically left open the *question of future management* of the 1.5 million acre coastal plain of the 19 million acre Arctic National Wildlife Refuge because of the area's potentially enormous oil and gas resources. . . . I have selected as my preferred alternative, making available for consideration the entire Arctic Refuge coastal plain for oil and gas *leasing*. . . . The step-by-step environmental planning, review, and evaluation procedures included in a *leasing* program *provide the best opportunity for the Department to decide what areas to lease, based on the most accurate and advanced information available at each step of the leasing process*. Although the exact process depends upon the leasing program established by the Congress.[7] [italics added]

This document is the government's means of handling asymmetric information.[8] Needless to say, reports of this type are time consuming, costly, and do not account for the subjective costs at hand. Anderson and Leal (2001, p.76) point out that "although experts can assess the time-and-place-specific environmental and technical constraints, all costs ultimately are subjective and depend on the values that individuals place on resources." Within the market, any changes in the asymmetric information relevant to particular decisions (such as the use of ANWR) are quickly incorporated into the price. The market is much more efficient and capable of handling asymmetric information because prices directly reflect individuals' subjective values.

Hodel (1987) clearly emphasizes that market forces do not allocate ANWR resources. Rather, these decisions are (and will continue to be) riddled with bi-partisan opinions of government representatives far removed from the actual issue and lobbyists who advocate actions for their respective groups. Problems arise since none of these groups have to bear the full costs associated with their choices, always a recipe for disaster. This third-party intervention cannot make the best decisions based on "the most accurate and advanced information available"[9] because the land in question simply is not private property. When people make a decision concerning the use of their own property, they take into account many more alternatives than they do when supporting political determinations about the use of property owned by others.

Let us consider the ramifications of a government-leasing program. Whichever group is granted such rights will only be entitled to use that property for a specified number of years—not to own the property. Now, to be sure, there is nothing wrong with a lease. This form of contract appears all throughout modern commercial endeavor. It is imperative, however, that some responsible adult, economically speaking, be the leaseholder, and we have seen that government is constitutionally unable to play any such role. This is especially important in environmental issues, since long-term considerations play an important role in this context. No renter of a car concerns himself with an oil change. The detrimental effects of failure to engage in such a procedure will not become apparent until the car lease has long ago expired. Thus, the car rental company must concern itself with such long-term planning. If it cannot, there is wastage, precisely what occurs when government gets into the environmental leasing business (Lee, 2001, p. 218).

It might be argued that government leasing may not be effective if the lessees are subject to a contract for only say, ten years. For this has the potential to minimize the incentives of these particular renters to take into account long-term supply and demand. If an oil company is granted a ten-year lease on a tract of ANWR, who is to say that it will take into account the value of the property after the ten years of oil drilling? This, however, is but a superficial critique of the present policy; for there is nothing wrong, in principle, with the ancient and honorable practice of leasing. If company A owns land, and leases it to B, and if B "runs it into the ground," then A will suffer. The difficulty in the present case is that when government is in charge, if it fails to ensure that the lessee maintains the property in an economical manner, it suffers no financial reverses that will bankrupt it. The state can always make up any such losses through taxation.

When examining the management of ANWR in the case of government leasing versus full private property, we must focus attention on the three essential components of private property rights—that they be (1) clearly defined, (2) defendable, and (3) transferable.

The property in question involves not only the surface ownership rights but that for minerals as well. Currently these are two separate commodities. If one party owns the surface land, another party can own the mineral rights underneath the same land. "[Alaska] state law generally does not allow a surface owner to prevent a mineral rights owner from developing his or her assets" (Snyder and Shaw, 1995, p. 2). When the Federal Government wants to open a tract of land for leasing, both sides must be allowed to compete in the marketplace for both the surface and mineral property rights. The government currently restricts environmental groups under current leasing practices from even bidding on oil and gas leases—which are contracts conveying the mineral rights (Anderson and Leal, 2001, p. 85). If environmentalists want to protect the land by purchasing the rights to it, they must not be restricted to only being able to own surface rights. Thus, it is essential ANWR be well defined in terms of surface and mineral rights. That is, the government must cede to private enterprise not only the land surface but also the right to dig.[10]

For the most part, the U.S. judicial system is well established to provide remedies against violations of private property rights. Most infringements on private property by another party, such as trespass, can be successfully resolved in a court of law. The property owner can sue the violator and seek damages for their wrongful actions, and an injunction against further such incursions.

But there are egregious exceptions. Primarily among them is government ownership, which does not allow individual users of the land such rights and incentives. In addition, government ownership allows for special interests, such as in case of ANWR now under examination, to prevail in the event of any conflict of interest. Then, too, there is the doctrine of eminent domain, under which the state may seize private property, sometimes merely to transfer it to another private citizen, and either pay nothing at all, or what *it* determines to be "market value" (Epstein, 1985; www.mises.org/blog/archives /002009.asp).

Private property arrangements allocate resources efficiently by enabling markets to transmit information about various individuals' subjective valuations of those resources. Asymmetric information is then revealed through prices, and the problems associated with differences in subjective valuations can be mitigated in the market. People exercise more prudence and care when making decisions regarding their own property than when doing so with the property of others. Therefore, successful solutions to issues concerning environmental resource allocation are best achieved under these circumstances, as opposed to government ownership or leasing arrangements. The following illustration provides evidence that private property rights facilitate coordination of interests, which result in mutually advantageous agreements.

THE AUDUBON EXAMPLE

The Audubon Society privately owns and operates the Rainey Wildlife Sanctuary, a 26,000-acre preserve located at the edge of the Intracoastal Waterway and Vermillion Bay in Louisiana. Since the 1950s, the sanctuary has been pumping natural gas from the thirty-seven wells located within the preserved land, and by 1995 had gained more than $25 million in revenues from the recovery of resources (Snyder and Shaw, 1995, p. 1). Although not usually thought of in these terms, this is a private business. As such, it has incentives to find the balance between conservation and resource development. The Society allows drilling to occur within its sanctuary because it can personally determine the degree of environmentally sound practices used in extracting the natural gas, as well as, in good capitalist tradition, benefit from the resulting profits. The revenues gained from cooperation allow the Audubon society to expand their conservation efforts.[11]

Defining the property rights of ownership was a major issue with the Audubon Society in the case of the Rainey Sanctuary, as it is with the current leasing practices in the ANWR. The Audubon Society obtained the property through a donation by Grace Rainey Rogers in 1924, at which time she transferred the surface property rights (Snyder and Shaw, 1995, p. 2). She made the donation under the stipulation that the land was to be used for "wildlife sanctuary only," and therefore any exploration and drilling for natural resources would have violated this condition (Snyder and Shaw, 1995, p. 1). This did not pose a problem for the Audubon Society at the time because its only incentives were to preserve wildlife. The question of the mineral rights did not come into play for another twenty years.

In 1940, the discovery of natural gas in the area brought about the development of an extremely productive well neighboring the Sanctuary. Audubon went back to Grace Rainey Rogers asking for her to allow drilling on Rainey, and in turn she agreed to allow it and share the royalties 50/50 (Snyder and Shaw, 1995, p. 2). Two different accounts of what happened with the mineral property rights have arisen.

First, the National Audubon Society made an authorized statement in a 1991 *World Energy Council Journal* article essentially stating that Rogers had not completely transferred the mineral rights. Due to this property rights issue, Audubon was compelled to sign a lease allowing drilling because if they failed to do so, then the party with partial mineral rights could have drilled for oil by any means they chose—potentially in an environmentally unfriendly means (Snyder and Shaw, 1995, p. 2). Second, these authors uncovered more evidence to suggest that Rogers had *not* retained any mineral rights in the transfer of the property. Instead, the Audubon Society was prevented from drilling due to the "wildlife sanctuary only" stipulation (Snyder and Shaw, 1995, p. 2). The difference between these two accounts is simply a matter of image for the environmentalists. If you accept the first, the Audubon Society comes across as trying to prevent harmful drilling practices from being forced on the sanctuary by the partial mineral owners.[12] If you accept the second, the Audubon environmentalists are viewed as self-interested individuals looking to maximize the value of their land both environmentally and monetarily.

The Audubon Society takes an anti-drilling position when it comes to ANWR,[13] because if their green views prevail in the political sector, they capture the benefits of preserving a pristine environment with only minimal lobbying costs—far less than what ownership would entail. Audubon does not personally incur the cost of giving up revenues from energy development; rather they are spread over the entire population because it is government-owned property. If environmentalist groups were to own tracts of ANWR (assuming both the surface rights and the mineral rights), a replication of the Rainey Sanctuary seems quite likely.

In eleven Western states, the U.S. Government owns 265 million acres of land used by some 25,000 ranchers. Jimmy and Frances Goss are just one of these cattle ranching families whose livelihood depends on the cattle ranching business. The Goss's ancestors settled in the area in the 1800s and herded cattle over 100,000 acres of forest and grassland. The government disregarded their homesteading right when the area the family's cows once grazed turned into the Lincoln National Forest. The government now owns this land and charges them to use what under libertarian principle is rightfully theirs (www.fs.fed.us/r3/lincoln/).

Thank goodness for the Audubon Society and its Rainey Wildlife Sanctuary. This is a heaven-sent illustration of a rather complex issue. This example is "invaluable to illustrating the fact that there are alternative possibilities which can result in market transactions that are often never considered in bureaucratic and political decision-making. The ANWR debate has been continuously discussed as having only two options, drilling for oil or not drilling in order to preserve the wildlife. Rainey shows how the establishment of property rights can result in a profitable 'compromise' for both parties."[14]

CONCLUSION

What would happen if all of a sudden private property rights enlightenment took over in Washington DC? Legislation would be enacted to open ANWR to an auction of private bidders. The two main groups, oil companies and environmentalist groups, would eventually be given the chance to bid against each other in an attempt to purchase the property rights of both the surface and mineral rights that are the most valuable to each interest. In the political arena, any government decision favors one group at the expense of another, and thus there is always a win / lose or zero sum game situation. Privatization always results in a win/win or positive sum game situation due to the nature of voluntary trade. It moreover has the possibility of producing a symbiotic relationship between environmental concerns and oil resource recovery as the Audubon Society example illustrates.

NOTES

1. This is because the businessmen are far more beholden to the dollar vote of the citizens than are the politicians and bureaucrats to the political voters. The former occurs every day, on numerous occasions; the latter, once every four years, assuming

George Bush does not "postpone" them. The dollar vote may be focused narrowly, on a blue tie, or a paper clip. In the ballot box, one must vote for Bush or Kerry, even if neither is fully satisfactory.

2. A current illustration of this can be seen in Newsweek, November 11, 2002, pp. 26–30.

3. Red on the inside, green on the outside.

4. It might be claimed that the appropriate owner is the U.S. government, and that these decisions ought to be made, in a democracy, by majority vote. But by what right does this organization claim ownership of the land? Certainly not by homesteading (Block, 1990; Locke, 1948; Rothbard, 1973, p. 32), for it is stipulated that at least so far, no one has mixed his labor with this particular land.

5. For additional critiques of the doctrine of "existence value" see Nelson, 1997; Rosenthal and Nelson, 1992.

6. For example, if oil producers claimed "existence value" in pumping oil and conservationists claimed "existence value" in the pristine land, what mechanism do we have to determine who would receive the greatest benefits from the property? How can the merits of these "claims" be determined scientifically? They cannot be. The present authors hereby assert that claims of "existence value" boil down to interpersonal comparisons of utility which are not recognized as valid in economic argumentation (Rothbard, 1997). They are illogical, and cannot be recognized in scientific economic discourse.

7. http://www.r7.fws.gov/nwr/arctic/purposes.html

8. http://www.economist.com/research/Economics/alphabetic.cfm?TERM=ANTITRUST; http://www.nobel.se/economics/laureates/2001/public.html; http://www.people.virginia.edu/~cah2k/assyy2k.htm

9. http://www.r7.fws.gov/nwr/arctic/purposes.html

10. This is to be sharply distinguished from the ad coelum doctrine according to which once one owns the land surface these rights also apply to a decreasing sized cone down the center of the earth, and, indeed, to an increasing sized cone into the heavens (www.netvista.net/~hpb/cases/alevizo1.html; www.straightdope.com/classics/a5_136.html; www.mrsc.org/mc/courts/supreme/055wn2d/055wn2d0416.htm). For a critique of this doctrine, see Block (1996) and Block and Block (1998), and also citations on homesteading.

11. In sharp contrast, when oil or natural gas is found on the property of others, whether private or governmental, the Audubon Society is a staunch opponent of similar development (www.audubon.org/).

12. This is ironic because these partial mineral owners are the ones who donated the land in the first place. The heirs of these questionable rights even went as far as to reduce their percentage of royalties from 50 percent to 40 percent. See http://prb.datalane.net/sanctuary.htm and Snyder and Shaw (1995).

13. See National Audubon Society website www.audubon.org/

14. This is a quote from an unusually insightful referee of *This Journal*, in response to an earlier version of this chapter.

Chapter 10

Genetically Modified Foods

First, one must understand the meaning of the term "genetically engineered." Doug Sherman and Gregory Jaffe, co-directors of the Biotechnology Project at the Center for Science in the Public Interest, define *genetic engineering* as "removing genes from one organism—a plant, animal, or microbe—and transferring them to another." Most genes are simply codes, or blueprints, that tell a cell to make a protein. Furthermore, this transfer of genes gives the plant some desirable trait ("Genetically Engineered Foods: Are they safe?" Nutrition Action Healthletter, www.cspinet.org/nah/11_01/).

Sherman and Jaffe illustrate this sort of biotechnology with respect to tomatoes:

To make tomatoes grow bigger and faster, the gene controlling large fruit size in one plant might be put into a plant that grows fast, with the hope of ending up with a plant that grows fast with bigger fruit. If the tomato doesn't taste good, researchers might also look for a "taste" gene from another plant ("Genetically Engineered Foods: Are they safe?" Nutrition Action Healthletter, www.cspinet.org/nah/11_01/).

In addition, biotechnology scientists already have genetically engineered plants that require significantly less pesticide due to the integration of transforming genes from other plants that fend off different types of pests. Sherman and Jaffe explain that

> these company scientists continue developing fruit that can be picked ripe without becoming mushy, coffee that's naturally caffeine-free, soybeans that don't trigger allergic reactions and that contain more healthful omega-3 polyunsaturated fatty acids, golden rice (rice enhanced with vitamin A),[1] virus-resistant plants, insect-resistant plants, drought-resistant plants, and fruits and vegetables that contain more nutrients or possibly even vaccines.

One might assume that these companies would receive great support for further development and innovation in light of all the potential benefits; however, many critics exist. For example, Sherman and Jaffe acknowledge the enormous potential of genetically engineered crops but admit, "we'll never realize this potential unless we make sure farmers in developing countries have access to cheap or free genetically engineered seeds, that the crops don't harm the local environment, and that the foods are safe" ("Genetically Engineered Foods: Are they safe?" Nutrition Action Healthletter, www.cspinet.org/nah/11_01/). These critics of genetic engineering companies worry that "multinational corporations that sell the 'miracle seeds' are in business to make money, not feed the world" ("Genetically Engineered Foods: Are they safe?" Nutrition Action Healthletter, www.cspinet.org/nah/11_01/)—and these critics are right.

IMPLICATIONS

Companies invest in research and development because they hope to acquire returns in the future. Entrepreneurs are motivated by self-interest in the market sphere rather than altruism but this does not imply that the entire economy does not benefit.[2] However, when critics argue that this technology should be handed out for free, they ignore the very basis behind the entrepreneurial spirit, which fosters technology and innovation. Adam Smith's (1776) "invisible hand" demonstrates that trade is not a zero-sum game as these critics claim.[3] Although altruism does not represent the fundamental motivation behind genetic engineering, this technology can deliver better crops with higher yields at a lower cost, which may further assist in feeding the world. When these companies sell their "miracle seeds" they exchange their product for a monetary price, namely revenue, and the consumers exchange their money for these products. Neither the producers nor the consumers would voluntarily exchange their property rights, if both did not feel that they would benefit from the exchange. The very act of voluntary exchange illustrates that each party gives up something because they value the item received in exchange greater than the item given up; thus, both parties benefit. Hence, voluntary exchange denotes a positive sum game.

Related to this argument against multinational corporations that control the seed supply, critics argue that genetically engineered crops exported from the Northern developed countries will worsen the provision of food. Critics, such as Greenpeace, argue that "sold at prices far below the cost of production, this export-driven policy has put local farmers in Southern countries out of business—making more people poorer and creating a dependence of food imported from the north" (Greenpeace Report, http://archive.greenpeace.org/~usa/reports/biodiversity/ftw.html).

However, this argument contains faulty logic. First, if corporations sold their genetically engineered seeds below the cost of production, they could never survive by incurring losses. Rather more likely these corporations sell their seeds, produced by economies of scale, at prices far below the cost of production in those underdeveloped countries. If this is the case, inhabitants of these underdeveloped countries should be thankful for the supply of an inexpensive or less expensive food supply. After all, if people suffer from hunger because they cannot afford to grow enough food to sustain themselves, the introduction of lower priced agricultural products will grant more people the ability to buy food and those who had the ability to buy sufficient amounts prior to the introduction of the inexpensive crops will now possess the ability to buy even more.[4]

Greenpeace assumes that poverty will not cease unless "more family farmers produce sustainable crops, not less. Bumper crops in the North have resulted in low prices, driving family farmers in the South out of business. Surpluses sent to countries in need of food are sold at ridiculously low prices, driving family farmers in those areas out of business" (Greenpeace Report, http://archive.greenpeace.org/~usa/reports/biodiversity/ftw.html).

If people are in dire need of food, how can an organization claim that the prices of the food supply are ridiculously low? It seems hypocritical to advocate higher food prices charged to starving people to support inefficient family farmers. The free market's fail-safe mechanism rewards efficient producers and penalizes inefficient producers. Therefore, if farming activities in underdeveloped countries entail less efficient methods of production at higher costs, these farmers should devote their time and energy to economic activities where they have a comparative advantage. While technology often receives condemnation because of its displacement of outdated industries and its workers, opponents fail to see the creation of a new industry and its need for new workers to maintain it. Furthermore, technology and innovation lead to more efficient, less costly methods of production, which therefore pass the savings onto the consumer via a lower price.

Further entwined with the criticism of multinational corporations controlling the seed supply is the criticism that too little regulation oversees these industries and therefore these industries may produce unsafe crops. However, entrepreneurs are guided by the market's fail-safe mechanism, where competent entrepreneurs receive compensation via increased wealth and incompetent entrepreneurs, those who make bad decisions and misallocate resources, receive penalties via losses until ultimately that inefficient firm is driven out of the market. Therefore, companies have every incentive[5] to produce safe, genetically engineered crops.

However, the article "Better dead than GM-fed?" of *The Economist* illustrates the skepticism of desperately poor countries to accept genetically engineered crops. Although 14.5 million people in Southern Africa are dangerously hungry and many have been reduced to eating wild leaves and pig food, Southern Africans refuse to import genetically modified food. *The Economist* depicts the irony of the situation: "Americans have been chomping GM maize and soybeans for seven years, without detectable harm. And compared with the clear and immediate danger posed by malnutrition, the possibility of being poisoned by frankencorn seems rather remote" ("Better Dead than GM Fed," www.economist.com/science/displayStory.cfm?story_id=133719).

Furthermore, when asked if we should be nervous about eating food that contains genes from other organisms, Sherman and Jaffe explain that often we aren't eating those genes due to the processing of the crop. This is illustrated by corn processed into corn oil or high fructose corn syrup. Moreover, most crops that we buy in the supermarket have undergone genetic alteration through breeding with wild relatives. However, this process of hybridization has occurred for thousands of years. While Gore admits, "This is not to say that plant selection is inherently dangerous; on the contrary, it is one of history's greatest innovations . . . indeed, plant breeding is almost as old as civilization itself." He regrets that "we have now taken the ancient process of seed and plant selection to a technological extreme" (Gore 2000, pp. 129–130). Sherman and Jaffe also add "this traditional cross-breeding often produces foods that contain genes and proteins that people have never been exposed to before. And, like it or not, we're constantly eating the genes and proteins of harmless bacteria that inadvertently end up on our food" ("Genetically Engineered Foods: Are They Safe?" Nutrition Action Healthletter www.cspinet.org/nah/11_01/). Ultimately, when an entrepreneur responds to an open niche in the market and invests into one's idea, that individual must incorporate all of the benefits and costs of one's decisions—included in those costs are safety issues.

Anti-market environmentalists also claim that genetically engineered crops will lead to greater environmental pollution. Because they view this possible outcome as a market failure, these opponents are quick to advocate government regulation or elimination of the genetic engineering industry. Although Saign admits "it is unknown whether biotech animals and plants, if released into nature, would alter the environment or food chains in a way that is detrimental to humans," (37) he still argues that government should create "an independent interdisciplinary ethics advisory board for the biotech industry, increase support for small and mid-size farmers, expand subsidies for organic and sustainable farmers, mandate that biotech food products receive long-term health testing, and ban all pesticides and herbicides until they are tested for long-term effects" (41).

However, these critics fail to realize that what they deem as market failures in reality constitute government failures. Neither individuals nor corporations should invade the private property of others. Machan asserts that "the tragedy of the commons arising from environmental pollution can be translated as the failure to extend the concept of rights into new domains" (164). Therefore, if the government correctly upheld private property rights, both individuals and corporations would have to incorporate the cost of pollution into their decisions, where this cost would include both damages and injunctions. Yandle adds, "Why not hold the polluter responsible for common law violation of adjoining environmental rights and either shut the polluter down or make the polluter pay damages?" (31)[6]

Advocating government regulation through the political process disregards the fact that politicians and bureaucrats too are self-interested, they receive rewards through responding to political pressure groups and expanding their agency size and budgets. Moreover, arguing for a distant third party to manage environmental issues neglects the fact that those individuals concerned with the protection of their own private property rights are more often better equipped to make decisions that coincide with good resource stewardship. Benjamin adds:

> It is tempting to conclude, as some have done, that because people may be prone to error in risk assessment, experts should make assessments on behalf of people who would otherwise err. . . . I would argue that only people who bear the consequences of decisions can fully know the advantages and disadvantages of each expert decision (Benjamin, 224).

We must admit that not only do experts err, disagree with each other, and that evidence is sometimes ambiguous, but also that experts who work for bureaucracies "have the same goals as the rest of us, they want higher incomes, promotions, discretionary authority, security, and a variety of work place amenities" (Johnson, 217).

Many anti-market environmentalists view technical innovation as the cause of our environmental and natural resource problems; however, Orr sheds light on the actual nature of technology—its neutrality: "If technology goes astray, it is because society gives the wrong signals to its managers about what it wants to conserve. If the signal is that environmental resources are free, we should not be surprised to find a technical response that leads to their profligate use" (Orr, 50).[7]

With respect to monoculture, anti-market environmentalists fear that genetic engineering will eliminate biodiversity. Therefore, just as Sagan advocates the creation of bureaucracies to regulate this industry, Gore expresses his concerns by stating "when we intervene in the process of evolution by directing the selection of those genetic characteristics that will be passed on from one generation to the next, the choices are usually based on the maximum yield and current market value of the varieties in question rather than their overall genetic resilience" (Gore, 129).

However, Gore ignores the fact that entrepreneurs do not concern themselves solely with the current market value of their investments, but rather the present discounted value, which incorporates all future revenues to be generated by the investment. Therefore, entrepreneurs will assess the risk of blight, drought, and pests devastating their investments. Moreover, our market system, which encourages specialization and division of labor, provides us with greater diversity than ever imaginable in the absence of a free market system. Likewise, Gore notes, "Most of us no longer produce our own food, but rely instead on a huge and complex apparatus that places an amazing variety of foods from every corner of the world in our supermarkets" (126). Through the free market system, an individual decides for oneself what to specialize in and through the coordination of markets the invisible hand leads to diversification.

CONCLUDING REMARKS

Although this chapter has supported the genetic engineering industry, it does not neglect the importance of choice. Therefore, if segments of the public prefer organic crops to genetically engineered crops, then the market will ensure that they receive organic crops for "profits link self-interest with good resource management by attracting entrepreneurs to open niches"(Anderson and Leal, 6). However, to advocate government regulation or elimination of an industry, because certain interest groups dislike the nature of a certain industry, interferes with the choice of others. Machan points out that "a crucial aspect of the thesis that human beings are moral agents is that human beings are able to make fundamental choices in life; that they possess free will and are the ultimate cause of their most important actions" (155). Therefore, the free market system, unhindered, will allow those individuals who would like to buy cheaper, genetically engineered food, to do so, while others who detest the idea can opt to buy organic foods.

Margarine coloring laws are mean-spirited and underhanded.

It is an old, old story.

When automobiles were first introduced, the horse and buggy industry was instrumental in obtaining all sorts of laws interfering with this new industry. For example, legislative enactments required people to run ahead of automobiles with flags or lanterns, warning all and sundry of the subsequent passage of a "horseless carriage."

Similarly, when supermarkets first came about, small grocers' associations attempted to hem them in with a welter of legal restrictions. These laws were so powerful in some countries (e.g., Japan) that this modern mode of shopping has still not fully taken root. Although attempts have been made to interpret such "restrictive entry" legislation as being in the public interest, all such claims are immediately and totally transparent.

And so it is with the present laws in Ontario and Quebec which restrict margarine coloration. They are blatant examples of special interest legislation, and should be seen as such, and repealed forthwith.

A case in point is section 4 of Ontario's Oleomargarine Act and similar regulation under the Dairy Products and Dairy Products Substitutes Act in Quebec. These laws interfere with the right of the manufacturers of margarine to color it in a manner compatible with consumer tastes; in a manner, that is, which will maximize profits and lead to a proper allocation of resources. Namely, the laws prohibit manufacturers from coloring margarine yellow, on the ground that the public might confuse it with butter.

Such restrictions might conceivably be justified under the general argument against fraud if there were even the slightest danger that the consumer would confuse the two products. But as these items are clearly marked, there is no case whatsoever for color legislation on these grounds.

What, then, is the purpose of these Acts? Adam Smith explained it best in his *The Wealth of Nations*, 1776: "People of the same trade seldom meet together, even for diversion or merriment, but the conversation ends in a conspiracy against the public." The butter, dairy, and other agricultural interests, unable to compete with the sellers of oleomargarine through the marketplace, have turned to the government for special privileges and protections. But the proper function of the state is not to discriminate in favor of any one special business interest or against another. Rather, its purpose is—or should be—to legislate in the public interest and ensure that no one can interfere with consumer sovereignty.

The margarine coloring restrictions are incompatible with Canadian anti-combines, anti-monopoly, and anti-conspiracy legislation. In addition, they impede free consumer choice.

The laws in Ontario and Quebec are reminiscent of Frederic Bastiat's story of the seventeenth-century French candle makers who sought tariffs to protect them against "unfair competition" from suppliers from other countries. Bastiat argued that they might, with equal logic, seek injunctive relief from the competition from the sun.

In the modern version of this tale, margarine makers represent the "big, foreign supplier" providing access to a "cheaper," more competitive product.

Oleo coloring restriction is mean-spirited and underhanded. Its ultimate effect is to single out, restrict, and punish an entire industry. If this is deemed to be appropriate public policy—not that any rational argument has ever been vouchsafed in its defense—at least this goal should be openly pursued. That is, instead of merely interfering with the basic human right of the margarine industry to use whatever color it deems advisable, government should subject oleo to an outright ban: treat it, that is, exactly as it does other illicit substances such as heroin, cocaine, and so on.

One benefit of such an "honest" policy would be to make clear to the public the logically implied result of the present oleo restrictions. Further, if margarine is banned, despite its healthy properties, and contrary to the wishes of the many citizens who are accustomed to consuming it, government would then have to pay the full political price for its action.

It is highly tragic that just when the Soviets seem to be relinquishing their seven-decades-long experiment with government intrusiveness into economic matters, that the Ontario and Quebec governments, through these restrictions, are remaining steadfast in their intervention in the economy.

Have we learned nothing from the failure of the Soviet and Eastern European experience? Shall we be condemned to follow their footsteps? One hopes not. Perhaps the first step that could be taken by Queens Park and the National Assembly in this regard is to repeal their unwise, unfair, and illogical interference in what would otherwise be a more competitive market between butter and margarine.

NOTES

1. According to Sherman and Jaffe (undated) "An estimated half-million children in the world go blind every year because their diets do not contain enough vitamin A. Millions more die from infectious diseases that their immune systems might have been able to fight off with enough vitamin A. By inserting two genes from a daffodil and one from a bacterium into rice plants, scientists have created rice with beta-carotene, which the body turns into vitamin A."

2. Leal and Anderson note that free market environmentalism is founded on certain visions regarding human nature, knowledge, and processes. With respect to human nature, free market environmentalism views man as self-interested. This self-interest may be enlightened to the extent that people are capable of setting aside their own well-being for close relatives and friends or that they may be conditioned by moral principles. But beyond this, good intentions will not suffice to produce good results. Instead of intentions, good resource stewardship depends on how well social institutions harness self-interest through individual incentives. See *Free Market Environmentalism*, pp. 4–5.

3. Ricardo (1821) is usually credited for the concept of comparative advantage.

4. A far more likely explanation of low food prices, a phenomenon that Greenpeace mistakenly blames on multinational corporations, is agricultural export subsidies in the United States and the EU. We owe this point to Glenn Fox.

5. The counterargument to this is the claim that this is the case only if decision makers are faced with all of the costs and benefits of their decisions. Opponents of GM believe there is a substantial risk from this technology not borne by decision makers. But surely "caveat emptor" puts paid to this notion. That is, all supposed externalities are internalized by that fact that there is a willing seller, and a willing buyer, where the latter bears the risk of any unforeseen developments. Were this not the case, then we would be at sea without a rudder: anyone could argue anything he wanted, with no limit. For example, perhaps the relevant costs of Frisbees are not taken into account. Who knows what evils lurk in the handling of this plastic toy?

6. It is sometimes asserted by opponents of free market environmentalism that genetically modified foods would lead to unwarranted spill over problems. The classical case of this is smoke pollution. But external diseconomies like pollution are evidence of *government* failure to uphold private property rights against encroaching distributors of trespassing smoke particles (Rothbard, 1990; Horwitz, 1977; McGee and Block, 1994). Nothing of the sort can even be imagined, let alone proven, in the case under discussion.

7. "Since many environmentalists believe that environmental amenities should be free to all, they support and promote programs intended to produce environmental amenities free of user charges. Furthermore, when the prices of environmental goods are set at zero, a certain dynamic process sets in motion. The public believes the environmental goods are free and, as a consequence, rationally demands that still more goods be produced. Why not, it's free?" The Environmental Costs of Bureaucratic Governance: Theory and Cases, by Johnson, Bruce of Baden and Stroup's *Bureaucracy vs. Environment*, p. 221.

Chapter 11

Economics of the Environment

THE HANDICAPPED AND THE ENVIRONMENT

If ever there were two groups that should have sympathy for one another—at least according to traditional political-economic analysis—it would be the environmentalists and the handicapped. After all, the leaders of each are strong proponents of the modern welfare-state mixed-economy philosophy. Both are accorded favored minority status, and are hence able to claim rights which would otherwise not be granted them.

Recently, however, they have been more at one another's throats than in each other's arms. Their bone of contention? Backpackers and the disabled are now waging a furious battle over the upgrading of an access road to Flat Mountain Pond. This body of water, located in a remote part of the White Mountain National Forest in New Hampshire, is not at all wheelchair accessible. Indeed, the 7-mile stretch of trail presents a formidable challenge even to non-handicapped persons.

The confrontation between the two groups has arisen over government plans to "upgrade" the trail by constructing thirteen rustic bridges. This would allow those confined to a wheelchair to be carried by automobile to the edge of the Sandwich Range Wilderness.

But this is the last thing the purist backpackers desire. For them, the unsullied forest is the ideal. It is the absence of all man-made creations —vehicles, bridges, and so on—which is why they seek the solitude of the woods in the first place.

However, they do not have the courage of their convictions. They could forthrightly include wheelchairs—and wheelchair-bound people—as one of the elements of modern industrial living from which they are attempting to escape. But this would be too confrontational. Who wants, after all, to take on the handicapped? Instead, they claim this particular plan is ill-conceived, and won't really benefit many people in wheelchairs anyway.

But there is a limit to prevarication. The tree-hugger philosophy is fundamentally incompatible with making provision for wheelchair accessibility. This is because an essential tenet of their view is that nature must remain just as is, untouched by meddlesome human hands. It is not that this particular plan is unsatisfactory. Any method which would gain for the handicapped access to remote parts of the forest must of necessity require interference with the natural order of things. Honesty and rigorous analysis compels us to recognize that this dispute is intrinsic, not just superficial.

How, then, can the battle be resolved? Is every bit of the world, is every inch of nature, to be paved, or bridged, or otherwise developed? Are the nature freaks to have nothing they can call their own?

Under the philosophy of positive rights currently in vogue, that is indeed the inescapable conclusion. For in this view, strictly speaking, all people have a right to go anywhere they please, and this applies with special force to the weakest members of society: women, natives, visible minorities, and particularly the handicapped.

There is, however, an entirely different philosophy, which could provide some measure of relief for the naturo backpackers. In this alternative classical liberal perspective, private property rights are a legitimate aspect of the just society. That is, there should be a private realm within which individuals may order their lives exactly as they please, despite the wishes of others to the contrary; and, further and far more controversial, that this realm of individual rights ought to include disadvantaged people such as the handicapped, and ought to be extended as far as possible.

What this means in the present context is that the forests, parks, and wooded areas of the nation ought to be privatized. Some of this land, presumably, would come to be owned by entrepreneurs interested in operating on an extensive basis: attracting clientele from all walks of life, including the handicapped. They would certainly develop the forest preserves so that all people could utilize them.

Other owners, however, would specialize. They might prohibit any and all interferences with nature, thus constituting a backpacker's haven. Or, they might limit themselves to just the handicapped, thus providing a refuge for them alone.

It is important to realize that the private property elements of the free enterprise system are the only institutional arrangements compatible with the complete safeguarding of at least some land in its natural untrammeled condition. The human rights philosophy cannot achieve this aim, as there are literally no places from which the handicapped (and other certified victim groups) can legitimately be excluded.

Yet it is one of the ironies of the modern day that even though the market is the last, best, and perhaps only hope of the ecologically oriented backpackers, they are biased against it. This is because many of the people in this community unfortunately believe that there is a basic incompatibility between the demands of economics and their own environmental concerns.

Perhaps the best intellectual antidote to this mistaken opinion is to realize that it is possible to reconcile the desires of many (but not all) people concerned with the environment, on the one hand, and on the other those who wish to defend private property rights and the free market. The reconciliation? It is that the economic means—prices, property, profits—can be used so as to attain the goals of protecting the environment. And not only that the market can be used as a means of protecting the ecological sphere, but that it is the best conceivable way of accomplishing this task.

THE GAIA HYPOTHESIS

The Gaia hypothesis is an enthusiasm that seems to have afflicted the legions of ex-flower children, hippies, the Woodstock generation, tree huggers, and other right (i.e., left) thinking people. In this view, it is not only true that animals have rights we are bound to respect as fellow sentient beings, but so do plants and vegetables, insofar as they share life forces with us. Even the very earth must be safeguarded from the rapacious designs upon it on the part of *Homo sapiens*.

The planet itself is seen as in some sense a living thing, a part—along with human beings, fauna, and flora—of some gigantic "being." As poetry, this may be a beautiful concept. But the advocates of Gaia mean it as much more than a work of art. Its devotees intend that it be used as an integral guide to public policy analysis.

If we take this hypothesis literally—and how else are we to take it if it is to be applied to the real world and its problems—Gaia is nonsense on stilts. The earth is not alive, contrary to the allegations of the Gaiaists. Our planet is not sentient; it is inanimate. It does not fit the usual definitions of life: ability to inject nourishment, eject waste products, and reproduce more of its own kind. We might as well say that the sun, the moon, the other planets, the stars, are alive. They are not, and to suppose that they are would be ludicrous.

The implications of Gaia are as troubling as is the hypothesis itself is false. Not only would mining be prohibited—how would you like it if someone came along and dug a hole in you and started taking things out of your body?—but so would plowing the earth. Strictly speaking, any interference with the earth—building homes, roads, parks, farms—would be illicit since the earth has "rights" to not be aggressed against.

Sustainable growth is the watchword of the Gaiaists. Population is permitted to grow and people can become wealthier, but only if this does not interfere with the "rights" of the other denizens of our environment, including the earth itself. What this means in practical terms, however, in the view of green central planners, is that population should decrease and our standard of living should fall to levels attained many years ago.

Part of the attractiveness of the sustainable growth idea is the implicit premise that if we exceed this rate the ecological system will boomerang on us, and we will have less prosperity than we would have had we not attempted this exercise in hubris.

But the marketplace, and the more traditional morality upon which it is often based, can achieve sustainable growth without the governmental interference required by the Gaia vision. According to the much more sensible Judeo–Christian teachings, the fish, beasts, and flowers of the earth, and the earth itself, are put here for man to enjoy and employ. This implication of the Bible, the main alternative to Gaia, does not mean that man can achieve optimal growth rates by sheer act of will or intelligence. No, the Gaiaists are correct at least in their assessment that these tools are far too puny.

Mankind need not rely on any human intelligence incorporated into centralized political planning of the economy. These methods have proven to be a dismal failure in Eastern Europe, the Soviet Union, China, Africa, and indeed everywhere else they have been tried. But the human race has a magnificent tool at its beck and call. This is the system of decentralized free enterprise, backed by a governmental commitment to enforce property rights and uphold the laws of trespass. This system can mobilize the bits of information held by millions of market participants. Through the price system, markets automatically guide human action.

Those who bite off more than they can chew, that is, attempt growth in excess of environmentally and economically sustainable levels, will suffer economic losses and even bankruptcy. If they chop down more of their privately owned forests than is optimal, they will destroy the present capitalized value of the future income streams they might otherwise have earned. If they pollute the atmosphere, they will be charged with trespass. If they invade lakes and streams with effluent, the private owners of these bodies of water will sue them for damages.

Sustainable growth is a reasonable goal, but this is equivalent to growth under free enterprise, and it needs no political intervention to guide it.

SUSTAINABLE GROWTH, SLOW GROWTH

East side, west side, all around the continent, the anti-growth forces are kicking up a fuss and making it next to impossible for the rest of us to get on with the business of making a better world for ourselves.

On Bowen Island, B.C., the West Vancouver Praxis Group announced plans for an 850-acre, $120 million residential-resort development. Over 200 acres of this Cowan's Point Resort would be allotted to recreational, wooded, and open space, while the remainder would be devoted to a 200-room guest hotel and lodge and to almost 400 homes on lots ranging from 0.5 to 2.5 acres. But spokesman for the Islands Trust—the local land use authority—has been cocking a snook at the proposal. Among the objections: it will interfere with the official community plans; place significant demands on the island's infrastructure; and necessitate upgraded ferry services.

There are many other efforts of this sort that have occurred in British Columbia. The Agricultural Land Reserve is a program designed to derail economic development and progress. Our zoning legislation makes it impossible to build high rises to world-class standards, imposing arbitrary and capricious height restrictions.

Petaluma, California, has instituted a myriad of restrictions on property development. This community—perhaps the world headquarters of the no-growth movement in that they have been at it for decades—has made it all but impossible for newcomers to enter the area. Their welter of restrictions, licenses, permits, and bureaucratic red tape are the envy of opponents of developers around the world.

Iowa City, too, has entered the fray. While small towns are usually presumed to roll out the red carpet for new businesses, this response is going the way of the dodo bird as the new anti-growth philosophy takes hold. In Iowa City, local voters have been creating difficulties for Wal-Mart, the third largest retailer in the United States and the fastest growing company in this field with $20 billion in sales. This chain had the temerity to propose construction of an 87,000-square-foot store 3 miles from the downtown area as part of a 28-acre suburban shopping mall. The company had anticipated $10 million in annual sales and scores of new jobs created, but an unholy alliance of ideological activists, competing downtown merchants, environmentalists, and others have been campaigning to stop any forward progress.

Wal-Mart was forced to withdraw its plans to open a similar emporium in several cities after activists protested.[1]

Similarly, the town council of Port Angeles, located on the Olympic Peninsula of Washington State, was able to derail a McDonald's attempt to place its golden arches in their community, in this case in the downtown area. Opponents cited the danger of Styrofoam containers, plastic wrappers, and increased traffic.

One of the problems with the philosophy of limited growth is that it turns the clock back to the medieval guild system. Under that regime, the privileged few were able to prohibit outsiders from competing. How else are we to interpret these savage and gratuitous commercial attacks on Wal-Mart, McDonald's, and the various resort developers?

But even more important, if this busybody tendency continues to gather strength, eventually there will be no private property rights left. How else can one characterize a situation where more and more firms are finding they must first obtain the permission of other people before they can do as they (otherwise legally) wish with their own possessions?

This way, however, lies in the direction from which the Soviets and Eastern Europeans are even now attempting to escape. Let us hope that Canadian citizens will insist that we not follow them down that particular dead-end garden path.

These new slow-growth no-growth initiatives are widespread, covering not only North America but the wealthy countries of Europe as well. In fact, it seems to have spread to virtually all the highly industrially developed countries in the world. But there is an anomaly here. Had this philosophy been in operation for the past century, there is great doubt that these nations would now be enjoying their present standards of living. Happily, and understandably so, this movement has not reached the Third World. They are keenly interested in improving their economic lot and would have done so long ago had it not been for their socialistic policies.

Nor is it democratic. The people who vote in favor of building restrictions are the present inhabitants of the local communities. Those who would like to live there and would be able to do so but for the multitudinous constraints are never heard from in the ballot box.

THE IVORY IMPORT BAN

When Daniel arap Moi of Kenya burned more than 2,500 elephant tusks worth an estimated $3.6 million in July 1989, he set off a conflagration that reached far beyond the ivory trade. His action had implications for environmentalism in general, for free trade, and for the privatization movement that is now sweeping the globe.

Moi set off the blaze in Nairobi National Park to dramatize the plight of the African elephant. Poachers have been killing these gigantic beasts at such a furious clip that there is now the fear in some quarters that the species may soon become extinct.

As a result, the price of ivory has been driven higher and higher. This so-called "white gold" currently sells for some $100 per pound. Unless something is done, the lure of profits will undoubtedly lead to still more intensive poaching, even higher ivory prices, and greater danger to the survival of the elephant. Moi set his bonfire with the hope that a worldwide ban on the importation of ivory would reverse this cycle. A trade ban that reduced demand would also lower ivory prices. With softer markets, there would be less incentive for poachers to operate.

The bonfire was clearly a public relations success. Even commentators generally receptive to the case for capitalism, privatization, and free trade have gone along with the idea as the only possible solution in this case.

Unfortunately, commercial bans and import restrictions simply do not work. Nor can one claim that all of the ivory now traded is the result of theft, and that it is therefore legitimate to prohibit trading in poached ivory. According to the estimates of the conservationists themselves, only some 80 percent of the ivory traded today stems from poaching. There are large stocks of ivory already stockpiled, including unworked material as well as extant jewelry, sculptures, piano keys, billiard balls, chop sticks, and Asian signature stamps. Some of this wealth may have emanated from poaching, but most, especially that created in the past, likely has not. In addition, not all African nations have gone along with the ban proposed by the Convention on International Trade in Endangered Species. One cannot therefore assume that if an object is made of ivory, it must be the product of poaching. Unfortunately, a trade ban covers stolen as well as legitimate property.

There is an alternative that would provide more sure protection for elephants without devaluing legitimate property. That alternative is privatization.

Privately owned animals are cared for and protected by their owners. The threat of extinction has never been even a remote threat for cows, sheep, goats, horses, dogs, or cats. Why, then, not apply this lesson to the African elephant? The first response is that elephants are different from cows; they need much more space and huge amounts of food. What remain of the wild elephant herds range over hundreds of square miles, often failing to recognize the sanctity of national boundaries.

A requirement for wide-open spaces does not seriously undermine the case for privatization, however. In the American West of the nineteenth century, in the epoch before barbed wire, ranchers would brand their cattle and then let them run loose on the wide-open prairie. Even in this setting, ranchers who shared the common pasture actively guarded against cattle thieves. A similar approach could help protect elephants in Africa.

In addition, the fact that publicly owned elephants have been allowed to range over huge spaces does not necessarily make that the optimal arrangement. Elephants are at risk precisely because the size of the areas over which they have traditionally roamed makes it difficult for the public sector to safeguard against human predators. Other options do exist. Private owners might follow the pattern set by a farmer in Texas who shelters rhinos on 80 acres of his 2,200-acre ranch. In addition, the Nature Conservancy has established a large system of private nature preserves on more than 1.5 million acres in more than a 1,000 locations. Further, numerous zoos keep elephants and other large animals healthy on far less terrain than that commonly used in Africa — not in small cages, but in large natural habitats varying from 100 to 700 acres. Instead of picturing the commercialized, privatized elephant unhappily hemmed in and limited to a small stall or bam, one can envision a wide open "Disneyland" for elephants.

Adequate food supplies should present less of a problem for privately owned elephants than for publicly owned animals. The ranchers, zookeepers, and private-sector environmentalists who own elephants will find it in their interest to provide the needed quantities of food to keep their elephants healthy.

Finally, there is the argument that it is extremely expensive, if not simply impossible, to fence in elephants. But this, too, fails. Zoos manage to keep their charges from wandering from their assigned habitats. Ranchers have been able to install metal fences strong enough to withstand the best efforts of a hard-charging 2,500-lb. bull rhinoceros named Macho: Private elephant entrepreneurs in Africa and elsewhere ought to be able to come up with something similar for their recalcitrant beasts.

It is difficult at first to imagine the elephant in the role of a barnyard animal because it does not give milk, lay eggs, or provide wool. Nor do we eat elephant meat. But the highly valuable tusks alone, to say nothing of the leather, would provide incentives to preserve and protect the elephant. Owners might also exploit tourists' interest in elephants through picture-taking safaris, controlled hunting, and sightseeing as a further means of paying for the animals' upkeep.

Let us consider one last objection. It is claimed that the poaching problem is so serious that unless something is done immediately, no elephants will be left to be privatized. Meanwhile, the process of turning over the elephants and the land they inhabit to private enterprise in Africa would take months, if not years. Therefore, opponents conclude that although privatization might be a good idea in theory, it will not work in practice.

In the first place, a ban on the ivory trade will also take months, if not years, for its effects to be felt. Even more important, this argument concedes in effect that privatization is desirable, but because it cannot be carried out instantaneously, the state must remain in control. On such grounds, one could object to *all* cases of deregulation and privatization. The fact that reform may take time does not imply that we should forego a superior solution rather than undertake the effort necessary to ensure the survival of the elephant.

Economically, the case for elephant privatization is airtight, and it could be accomplished in a matter of weeks if the political will were there. All that needs to be done is to set up an auction of, say, 100-elephant, 10-square-mile parcels. If this is not the optimal configuration, market trading will move the economy in the necessary direction. By all means, let the government do what it can in the meantime to preserve the elephant, but the best long-run solution is still privatization.

None of the arguments against elephant privatization can withstand close scrutiny. Despite pursuing a worthwhile goal, Daniel arap Moi and the Convention on International Trade in Endangered Species are traveling a dead-end road in seeking a worldwide ban on the ivory trade. Indeed, an ivory ban could actually make the situation worse.

It is true that the African elephant's misfortune is its tusks, but only in the present public ownership situation. Under a system of private ownership, the elephants' tusks would become a blessing, as private owners would care for and protect their elephants and guarantee the survival of the species to ensure future profits from the sale of the ivory. Moi's tusk bonfire captured the imagination of the world. Now let us encourage adoption of the policies that will ensure the survival of the animals. Privatize the elephants.

A PLAN TO SAVE OUR ENDANGERED SPECIES

The latest outrage on the environmental front is the tragic and wholesale elimination of the elephant and rhinoceros in Africa. The indiscriminate hunting, poaching, and killing of these creatures is so severe that actual extinction is feared by many ecologists.

Interventionist ecologists wax eloquent about the needless extinction of species. In their view, the villains of this piece are profit, development, and excessive demands for consumer goods. Their solution? Cut down on excessive and frivolous consumption, and end our reliance on the private property and profit systems. Many such environmentalists have called for a ban on the importation of elephant ivory, used for works of art, and the rhinoceros horn, used as an aphrodisiac, in an attempt to undermine the market for these products.

We do owe a debt of gratitude to these commentators for pointing out the perils to these large mammals. On the other hand, we cannot thank them for their insights as to the cause of the problem or its likely solution. In fact, the reason many species are brought almost to the edge of extinction and some, unfortunately, past this point is usually for the opposite reasons of those suggested by these ecologists. The problem is not an excessive reliance on private property and the private sector, but rather a lack of private property rights and the natural actions of the private sector.

Consider the buffalo and the cow. Biologically, they are very similar animals. And yet it is only fortuitous that the buffalo, which for many years was allowed to run free, unowned by man, was saved from extinction. In sharp contrast, the cow has been domesticated for millennia; it has been owned and thus cared for by farmers and herdsmen from biblical times and even before.

Economists call what almost happened to the buffalo—and what is currently happening to the rhinoceros and elephant—the tragedy of the commons. If no one is allowed private property rights over the buffalo, then it does not pay anyone to protect them or to see to it that they are not hunted to extinction. When a buffalo died in the non-private property days of "home, home on the range" and "don't fence me in," no one lost any money. As can be expected, no one acted to prevent such occurrences. When a cow dies, in contrast, the owner suffers.

Now consider the rhinoceros and elephant. The reason they are being driven to the vanishing point is not because of greed. It is because the countries of the world in which they can flourish have made it all but impossible for farmers to domesticate these animals and raise them in large herds for purposes of profit. There are no private interests aligned to support them, only bureaucrats intoning platitudes about the importance of refraining from hunting these beasts. Like bureaucrats the world over, these may be well-meaning people, but they have no real incentive to perpetuate the species. What is needed is a private, profit-seeking supplier responding to the evident demand.

It is sometimes said that the sheer value of these animals is what has proved to be their undoing. According to *The Economist*, ivory is worth $150 per kilogram; a single rhinoceros horn can sell for several thousands of dollars. But under a regime of private property, these evaluations would impart strong incentives to their owners to maintain and safeguard these animals.

Very valuable pelts are derived from the sable, mink, and chinchilla, and yet it is ludicrous to suppose that these domesticated animals are anywhere near extinction. Their owners have a financial incentive to maintain their investments. Even the lowly cow is worth several hundreds of dollars at the slaughterhouse. This is precisely what keeps that species alive and thriving, however.

These left-wing environmentalists have it backward. If they really wish to see the rhinoceros and elephant population prosper and grow, they should advocate private property, not reject this institution.

PETER PEARSE: A MODERN-DAY HENRY GEORGE

Henry George was an economist who achieved a measure of notoriety many decades ago. During his years of practice he gathered around him a movement of no small proportions. Dedicated to the principles of free enterprise, he waxed eloquent about the virtues of the market in allocating resources, promoting prosperity, and fighting poverty.

But there was one element of the economy, in his view, to which individual control ought not to apply: land. The profit and loss system could not be trusted to manage this resource as it could all others. So Henry George proposed a 100 percent tax on the value of land, thus effectively socializing it, while favoring the system of decentralism and economic competition for just about everything else.

Weird, you say? All right for the nineteenth century, perhaps, but laughable in the sophisticated twentieth? It can't happen here?

Well think again. One of British Columbia's eminent economists has just weighed in with an analysis which contains strong elements of the views put forth by Henry George many years ago. Speaking to an international forestry conference in Montreal, Professor Peter Pearse of the Forestry faculty of the University of British Columbia quite rightly criticized present-day forestry practice.

Government, not private interests, controls and manages our nation's woodlands, and this has led to their shrinkage and deterioration. The lumber firms have little economic incentive to replant. They are required to do so under commercial contracts with the state apparatus, but internally created profit and loss incentives are far more effective than inefficient bureaucrats who are only weakly dedicated to this task. Pearse notes that the forestry companies spend roughly four times as much replanting the small parcels of land they own outright as they do on acreage leased from the Crown. As well, they have little incentive to invest because, according to painful experience, they cannot legally rely on free farm licenses. They are all too likely to be revoked long before a tree can come to maturity.

Based on this analysis, Pearse courageously recommends the privatization of the forest resources of Canada. In his view, timber license arrangements should be replaced by timber rights. The forests should be sold to the lumber firms. However—shades of Henry George—the UBC professor fails to carry his insights forward to their logical conclusion. He calls for the privatization of forests but not of the land upon which they sit.

Privatization of the forests will rationalize the development of this resource, as we have seen, but, failure to privatize forest land will undermine the rational use of that very distinct resource. With continued Crown ownership, the allocation of land for wood growing versus other uses will be determined by the same political-economic system that doomed the Union of Soviet Socialist Republics, namely, socialism.

Suppose people come to value forest land in some regions more highly for other purposes, such as for farming or housing. Will they be able to bid these lands away from forestry to use for these other purposes under the Pearse scheme? Not a bit of it. Instead, they will have to use the political process to attain these essentially economic ends, just as was done in the bad old USSR. At a time when the people of Eastern Europe are embracing the market with alacrity, it is a shame and a disgrace to find a UBC professor counseling against a movement in this direction. Why even the relatively hard-line Gorbachev is now open to the concept of private ownership of land in Russia.

Cannot our local intellectuals at least keep up with the winds of change emanating from that part of the world?

Perhaps we do Professor Pearse an injustice. Might it be that his failure to carry through from forest privatization to forest land privatization is due to his concern for the well-known and well-established "preference for public ownership of the land" on the part of Canadians? If this is so, it is equally unconscionable. The duty of the public policy analyst is not to kowtow to the irrationalities of the community, if such be the case, watering down the implications of scientific analysis on their behalf, but rather it is his responsibility to honestly acquaint us with the results of his findings—let the chips fall where they may—and not to sugar coat the pill.

What would we think of the biochemist whose research indicated that smoking causes cancer, but, because of fears that the public will not accept this message, or out of concern for the reigning scientific paradigms, recommends that people cut down on their cigarette usage, but not quit outright? The same evaluation applies to those who know full well the many benefits of private property—economic, ethical, and efficiency—and yet refrain from acquainting the public with this insight.

NOTE

1. https://www.nytimes.com/1993/10/21/us/in-2-towns-main-street-fights-off-wal-mart.html

Chapter 12

Pollution

CLICHÉS OF POLITICS

"Private enterprise leads to pollution."

Before we can determine whether or not this statement is true, we have to define our terms.

Pollution is relatively easy. This refers to a private property border crossing: waste matter is transported from one person's holdings to that of another, either directly or indirectly. It is important that we interpret pollution in this manner; otherwise, it can easily be confused with something that superficially resembles it but is actually quite different, namely waste disposal. In the latter case, the byproducts of the processing of goods or services are retained by the firm which created them. It disposes of them without negatively impacting others' rights. In this case there is no "spillover" effect nor "externalities" affecting third parties.

The definition we are considering here, then, is a legal definition, not a physical one. As far as chemical analysis is concerned, it is impossible to distinguish pollution from waste disposal. In both cases, the waste material may well consist of identical chemical compounds.

The simplest example of pollution using this terminology is when Mr. Jones dumps his garbage on Mr. Smith's front lawn. That would be a case of direct pollution. The indirect version occurs when there is an intermediary between Jones and Smith, such as air or water. Here, Jones first incinerates his refuse, and prevailing winds carry the same material over to Smith, his lawn, and his lungs. Alternatively, Jones could dump this material into a river, negatively impacting Smith who lives downstream and is forced to use the now polluted water for drinking, irrigation, and washing.

Non-polluting waste disposal occurs when baseball fans leave the stadium littered with peanut shells and beer cans. There is no "border crossing" in this case, since ticket prices reflect the cleanup costs. The management voluntarily disposes of the detritus. Another example would be hiring a maid to clean one's home. There may be garbage strewn all over the place, but this is not pollution; the maid has agreed to do the cleaning for the owners.

What, then, is a private enterprise? This is not a system where government grants special privileges (subsidies, protection from competition) to business concerns. Far from it. On the contrary, free enterprise is a system based firmly on the premise of private property rights. Trespassing and theft are absolutely incompatible with the free market system. The market is the concatenation of all voluntary trades (employment, barter, purchase, rental, sale, etc.) between mutually consenting parties. Under pure and unadulterated laissez-faire capitalism, the only function of government is to protect person and property against physical attack from outsiders.

We are now ready to put these two elements together. Pollution, we have seen, is in effect a bombardment, an infringement of one person's rights by another. But the essence of private enterprise is precisely to prevent such goings-on. It can therefore be stated, without fear of contradiction that private enterprise does not lead to pollution. But this way of putting matters is far too understated. Actually, "market pollution" is a contradiction in terms. To the extent that there is private enterprise, there cannot be pollution; it is against the law. To the extent that pollution exists, this is prima facie evidence that the legal underpinnings of capitalism have been breached; that the law of trespass is not being upheld; that private property rights are not being defended. Pollution is a trespass or border crossing; the sole function of a free enterprise government is to prevent this (and other) violations of property rights. Therefore, the two are entirely incompatible.

Now it is of course true that many countries, in many areas of the world, in many time periods, have been considered to be based on free enterprise principles, and yet suffered from pollution. This only attests to the inaccuracy of language.

Let us consider one such example, the United States in the nineteenth century. Up until approximately 1830, the courts had based their decisions in "nuisance cases" (we would now call them environmental litigation) on a reasonably close approximation to a free enterprise legal system. If a farmer could show that the railroad engine was spewing forth sparks and setting his haystacks on fire, he could collect damages. If the housekeeper complained that factory fumes were dirtying the clean laundry she hung on her clothesline, she would typically be granted a cease-and-desist order. Injunctions were invariably granted to downstream users victimized by upstream waste dumping.

Under these conditions, it was clear that manufacturers had to take into account third-party effects. There were strong incentives to use cleaner burning but more expensive anthracite coal instead of the dirtier but cheaper high sulfur variety; to install smoke prevention devices; to engage in research and development aimed at abating nuisances. There was even an "environmental forensics" industry in the making, dedicated to determining guilt for pollution. And when all else failed, the railroad, the factory owner, and the "upstreamer" could pay their victims to the latter's satisfaction, thus converting pollution into waste disposal, and "internalizing" the externalities.

But in the 1850s and thereafter, a new philosophy began to permeate the legal fraternity. It was determined that the "public good" required economic progress. In the view of an increasing preponderance of judges, this could only be attained by supporting manufacturing. So when the aggrieved victim of pollution next appeared before the bench, they said, in effect, "Our primary goal is to facilitate a rising GNP. In order to do so, we must give carte blanche to polluters. Your selfish private property rights are in the way of the greater good for the greater number, and must be swept aside."

Under this condition, all market-oriented environmental incentives came to an abrupt halt. Previously, environmentally sound acts resulted from both selfishness and benevolence. Now, only the latter could operate, the one that Adam Smith saw as far less reliable. After all, why should the profit-seeking firm use clean fuel, or worry about smoke prevention, when it was not legally responsible for damages? The few businessmen who did so in any case, perhaps on moral grounds, put themselves at a competitive disadvantage; they became more liable to bankruptcy.

The resulting environmental crisis was due not to free enterprise, but to its very opposite.

EXTERNALITIES AND THE ENVIRONMENT

We operate under a free enterprise economic system that produces plastic milk jugs and redwood picnic tables. The market is therefore responsible for such environmental problems as too much plastic trash and too little conservation of forests. So goes a common belief, anyway. In fact, it is governmental failure to maintain and defend the institutions of a free market that is responsible for the environmental damage caused by private businesses.

In the case of solid-waste management, plastic companies and their customers escape from the cost of disposing of plastic after the consumer is finished with it.[1] This is because most garbage collection is organized through the public sector. The cost of disposing of the plastic and the other waste is undertaken by the government, and a citizen is typically taxed without regard to the amount of trash he generates. Once the citizen pays his taxes, he has no incentive to choose environmentally sound goods because disposal costs are in effect free to him.

If, instead, there were complete privatization of the garbage disposal industry, those who generate trash would directly pay for disposal costs. The owner of a private dump tends to charge tipping fees that vary with different kinds of trash. The price will be significantly higher for material that creates toxic waste[2] because the dump owner will be liable for any harmful leaks from his site. The hauling firm, which collects the garbage from the homeowner and must pay the tipping fee, will pass the price onto consumers. Consumers, knowing that they will have to pay more for the disposal of more plastic, will tend to substitute toward less costly, and thus more environmentally sound, containers. In the jargon of economists, the negative externality will disappear; the cost of trash disposal will be internalized, brought to bear on the responsible parties.

It is clear from this example that the negative externality currently related to solid-waste management is a governmental failure, rather than a market failure. By not allowing the free market to operate, the government pushes the costs of waste disposal onto the taxpayers. In a free market the price system accurately enables us to compare resources to determine which courses of actions are most economically and ecologically sound.[3]

Critics argue that the market fails to consider environmental concerns. In *Making Peace With the Planet*, Barry Commoner insists that the free market system conflicts with a social concern for environmental quality, and thus argues that ecologically sound production decisions must be implemented through planning.[4] Commoner fails to realize that when we turn away from the market, we are unable to compare resources and their values. Nor does he realize that the incentive structure in the political sector fosters an economically unsound approach to the environment.

The distorted incentives that operate in the public sector are responsible for many of our forest removal and logging problems. The difficulty stems from the fact that 42 percent of all U.S. land is owned by government.[5] The public forestry services do not operate by maximizing the value of their resources. Rather, they subsidize special interest groups who in turn support public ownership. For example, the U.S. Bureau of Land Management (BLM) uses a method known as chaining, which uproots trees, leaving holes and unsightly tracks, to remove trees from grazing lands. Although this costly method is seldom practiced in the private sector, the BLM has no incentive to minimize costs. It maximizes its budget by providing subsidized grazing rights for ranchers who in turn lobby for BLM expenditures.[6] The U.S. Forestry Service subsidizes logging companies to cut down trees on public lands by building logging roads for them.[7] The private firms thereby shift some of the costs of logging onto the public. Again, prices do not reflect the full costs of environmental destruction. Certainly if the business firms actually owned the land, they would better care for it because abusing it would reduce their prospects for future income.[8]

Government ownership of so much land artificially encourages logging for another reason as well. Private firms with large landholdings are discouraged from using them for hunting or recreational purposes; the government provides parks for these purposes at a zero or nominal price. In the national forest surrounding Yellowstone National Park, the Forest Service charges no user fee for elk hunting. This reduces the value of elk resources on private land and discourages private firms from devoting their forests to hunting rather than logging. The low level of private development of recreational land is likewise due to the minimal prices the federal government charges for the use of its recreational facilities.[9]

Unfortunately, because the federal government owns the land, environmental groups generally have to work through the political sector. They lobby to persuade the government to preserve land, in conflict with the opposing special interests of logging, ranching, and extraction industries. If the demand for environmental amenities were instead channeled through the marketplace, tremendous progress would be possible.

Voluntary trade allows for creative deals in which all parties gain, or else the trade does not take place. If current leasing arrangements were changed, environmental groups could bid to purchase or lease public resources.[10] On land that becomes their property, they can either preclude development entirely or sublease the land for development on their own terms. When natural gas was discovered in the national Audubon Society's Rainey Wildlife Sanctuary in Louisiana, for example, the group itself faced the tradeoff between strict preservation and drilling income, and struck a balance that allowed drilling under specified environmental conditions. The Audubon Society received royalties, which gave them the funding to purchase still more land for preservation.[11] In this case both the Consolidated Oil and Gas Company and the Audubon Society gained. Economic efficiency and environmental interests were both served.

The key to environmental protection is a free market with defendable and transferable property rights.

NOTES

1. William Rathje, Rubbish, *Atlantic* (December 1989), pp. 99–109, calls this assumption into question.

2. Walter Block, Resource Misallocation, Externalities, and Environmentalism: A U.S.-Canadian Analysis, *Proceedings of the Twenty-Fourth Annual Pacific Northwest Regional Economic Conference*, 1990, p. 93.

3. Walter Block, Environmental Problems, Private Property Rights Solutions, in Block, ed., *Economics and the Environment: A Reconciliation* (Vancouver, BC: Fraser Institute, 1990), p. 287.

4. Barry Commoner, *Making Peace with the Planet* (New York: The New Press, 1992), pp. 223–227.

5. Progressive Environmentalism Task Force Report, p. 27, National Center for Policy Analysis, Dallas, Tex., 1991.

6. John Baden, Crimes Against Nature: Public Funding of Environmental Destruction, *Policy Review*, Winter 1987, p. 38, cited in Walter Block, Environmental Problems, Private Rights Solutions, pp. 296–297.

7. Joseph L. Bast, Peter J. Hill, and Richard C. Rue, *Eco-Sanity* (Chicago: The Heartland Institute, 1994), p. 203.

8. Terry L. Anderson and Donald Leal, *Free Market Environmentalism* (San Francisco: Pacific Research Institute for Public Policy Research, 1991), p. 71.

9. Ibid., pp. 60–62.

10. Ibid., p. 93.

11. Ibid., p. 90.

Section VI

CLIMATE; NATURAL DISASTERS

Chapter 13

New Orleans' Future

Katrina and the Future of New Orleans

INTRODUCTION

There are no controlled experiments in economics, or, indeed, in any of the other social sciences. Social economic reality is simply too complex. Ceteris paribus conditions never hold; there are always several phenomena that vary, not the single one necessary for controlled experiments of the sort enjoyed in the physical sciences, where everything else can be held equal. Every once in a while we come close, however. In past decades there was East Germany and West Germany. North and South Korea serve similar functions at present. In each case, there was at least a rough approximation of the people, of the history, of the wealth, of the culture, and even of the language. Then, for historical reasons, each of these common groups was broken into two, whereupon they followed very different political economic philosophies, with very different results for the participants. Yes, it is still possible for those with closed minds to deny that communism is an inferior system when compared to the moderate version of capitalism practiced in West Germany and South Korea. They can still maintain "real" communism was not followed in East Germany and North Korea.[1] But, for most people, these semi-controlled experiments furnish(ed) a stark example of the results of these two very different systems.

Thanks to Hurricane Katrina and the failure of the socialist government levee system "protecting" New Orleans, we are possibly to witness a different kind of semi-controlled experiment. This one will be a time series controlled experiment, where we can view the Big Easy before and after Katrina and her aftermath; that is, if the advice of the present chapter is followed. To wit, we advocate a radical move toward private property, individual initiative, free markets, and laissez-faire capitalism. If these public policy recommendations are followed, we will then have a stark contrast between the aftermath New Orleans under its present socialist control, compared with something very, very different.

We approach this topic from a libertarian perspective. This might be seen as "bipartisan," as it renders us as critical of Democrats as of the Republicans. But it is not. It is very partisan, in that it advocates a completely different and competing system of social organization than either of these almost indistinguishable alternatives.

Libertarianism may be succinctly defined as being based on the motto "that government which governs least governs better, and that government that governs not at all is best of all." Libertarianism is predicated on two pillars: private property based on homesteading, and the non-aggression axiom. This means that justice in property titles consists of those, and only those acts, that can be traced back to voluntary economic interaction of rightful owners. This political economic philosophy is neither of the right nor the left but is rather something unique. It favors personal liberties; laws prohibiting victimless crimes concerning addictive drugs (heroin) or sex (prostitution) would be repealed. Paradoxically to some, it also supports economic liberties; laws prohibiting economic freedom (minimum wages, rent controls, tariffs, affirmative action, antitrust, price gouging) would similarly be ended.[2] Most important, the libertarian opposes foreign military adventurism; in this view, the United States would first off confine its military to the fifty states and no longer serve as the policeman to the world.[3]

BACKGROUND

Watching the Capitol Hill hearings on what went wrong after Hurricane Katrina provided a glimpse of what it must have been like in the Politburo in the 1950s. The Soviet bureaucrats would gather with the party officials and factory and farm managers to figure out why grain production was down or why shop shelves were empty or why the bread lines were ever longer and the quality ever worse.

They gathered under the conviction that they had a workable system that was being rendered unworkable because of the incompetence or shirking or wrecking of certain key players in the chain of command. No one was permitted to say that the command system itself was the problem. Instead, they had to place blame on someone, as if all problems could be reduced to issues of obedience. It was always a scramble. Whoever was finally said to be at fault faced certain ruin.

To be sure, there was plenty of blame to go around. As with scapegoats the world over, there is a sense in which they are all responsible for not having found the exit. If any of those scapegoats could also organize into a hierarchy of control and hold trials, it would surely produce a great show and many victims. But, they would be no closer to getting to the real source of their problems. And so it was in the U.S. Congress: the hearings produced a great show with no results that will make a difference for our future.

The Soviet system had to fully unravel before it became permissible to state what it used to be a crime even to think: an economy cannot be centrally commanded. The bureaucrat can make every demand, issue a million orders, exhaust every financial resource in the state's account, elevate some people and demote others, dress up in a military costume and make grand pronouncements from a glorified pedestal, cut off fingers, toes, and heads, but in the end, he cannot make the economy perform in a way that serves the people unless market forces are allowed to work.

Not just the Soviets[4] had to learn this. Authoritarian regimes from the beginning of time have attempted to defy the laws of economics, step on the interests of the merchant class, control and redirect the wishes of consumers and entrepreneurs, bend and kick prices and wages this way and that, and inhibit trade in every which way. But they cannot finally overpower the driving desire on the part of people to control their own fate and not be subject to the slavery that is collectivism of all colors, whether red or brown.

Someday, the U.S. managers of crises will have to realize this same point. But for now, they are like Soviet bureaucrats scrambling to make an unworkable system function, and creating a scene that is as farcical as it is tragic.

Consider first how the much-glorified Department of Homeland Security responded to the Katrina crisis. There is a mysterious missing day between the time the hurricane hit and the levees broke and flooded New Orleans. During this strange Monday, August 29, 2005, a day in which there was a window of opportunity to prevent the meltdown of civilization, why didn't federal officials respond or even pretend to do so?

The head of the Department of Homeland Security, Michael Chertoff, said that he read in the Tuesday morning newspaper that, according to the headline, "New Orleans Dodged the Bullet." So, to his mind, there was nothing to do. This was his testimony. This is not exactly an awe-inspiring admission, but it speaks to a truth that few are willing to admit: government officials live normal lives. They do not partake of the mind of God. They get their news the same way we all do. And they have far less information than the body of knowledge generated by the signaling process of the market economy and the private sector.

We might even say that they are in effect sub-normal in intelligence, because government officials stand outside of society, cut off from normal channels of information[5] that the rest of us take for granted. They are isolated from markets and the regular pressures of life. They are not owners of what they control, and have no real stake in the value of their product. They are surrounded by some of the most peculiar people in the world, namely lifetime bureaucrats, power-mad politicians, and lobbyists on the make. This is their world and this is what they know.

They enjoy the illusion of being better informed than the rest of us, so it would never occur to a high official to surf Google News to find out what is really going on. Thus, it was apparently beyond the capacity of FEMA to find out that the National Weather Service (NWS) had issued a flood warning soon after the hurricane hit. The NWS in turn was only reporting what many private local media outlets were saying.

Certainly the municipal government of New Orleans got the message. It issued a warning to residents, and then all the officials packed up and headed to Baton Rouge. We suppose this was the plan that the bureaucrats came up with after having received a $500,000 federal grant in 1997 to design a comprehensive plan for evacuation. Half a million dollars later, they agreed on what the plan should be. Two words: "Let's go!"

We can learn a great deal from observing all this. It is always the case that the government's first interest in a crisis is to protect itself. The "public interest" is way down the list. Its employees have no ancient code that requires them to go down with the ship. The seafaring captain might feel disgrace if he lost his crew and passengers but returned safely to shore,[6] but the bureaucrat would see this as nothing but rational self-interest at work.[7] From their point of view, public service is not a suicide pact.

If this is so, are we wise to expect government service in times of crisis? Well, here is where it gets complicated. They always promise that they will take care of us. On the day the hurricane hit, for example, President Bush made the following announcement: "For those of you, who are concerned about whether or not we're prepared to help, don't be. We are. We're in place. We've got equipment in place, supplies in place. And once the—once we're able to assess the damage, we'll be able to move in and help those good folks in the affected areas." Given the calamity that followed, this statement by Mr. Bush might as well have been a Soviet propaganda poster about the glorious future of socialism. If the only response by government had been to turn and run, they could be accused of hypocrisy, but it would have been better than the alternative of bad government that stayed to ruin the work that markets and private individuals do.

As the hurricane approached, for example, Mr. Bush, along with nearly every office holder in the entire region, immediately announced that there would be no tolerance of the so-called price gouging. What is and what is not gouging remain undefined by law, but there are still criminal penalties attached to doing it. If a merchant raises his prices to the point where he attracts a complaint, there is a pretty good chance that he will be thrashed as a gouger.

And yet, we have to ask ourselves what the purpose of a price is. It is a signaling device that allows market players, including both producers and consumers, to adjust their economic behavior in light of supply and demand. If supply remains the same and demand rises, the price too will have to rise so the market can clear properly. Otherwise there will be shortages and surpluses that will prove to be a benefit to no one. Bill Anderson[8] has called gouging rules a form of back-door price controls, and he is right. They create victims, encourage economic dislocations, and foster black markets.

One might think that a Republican administration would understand this, but reflect on the fact that Iraq still has very strict price controls on gasoline—controls[9] that were instituted by the United States after Saddam was overthrown. Don't think for a minute that it is beyond the capacity of the Bush administration to do what the Nixon administration did, which was to believe that the laws of markets can be overridden by regulatory force.[10] Anti-gouging laws, to the extent they are obeyed, will create shortages. But in telling the sad tale of Katrina, we would like to begin not with a case of shortage, but with a strange case of surplus.

One week after the hurricane, FEMA ordered the Army Corps of Engineers to buy 211 million pounds of ice from IAP Worldwide Services of Florida. Trucking companies were notified of a grand opportunity since the government was paying the bills for delivery, and the dispatchers sent out the word. There is no space to explore the workings of IAP Worldwide, but the company, which relies heavily on tax dollars as a federal contractor, has a new CEO who most recently held the position of vice president of national security programs for the notorious Kellogg Brown and Root. His name is David Swindle.

But back to the story of the ensuing chaos. One trucker picked up ice in Greenville, Pennsylvania, and was told to drive it to Carthage, Missouri. But when he arrived in Carthage, he was told by a FEMA official to go to Montgomery, Alabama. After a day and a half sitting in Montgomery, he was told to go to Camp Shelby, Mississippi, after which he was sent to Selma, Alabama, after which he was sent to Emporia, Virginia, where he stayed for a week burning fuel, until he was sent to North Carolina, and finally to Nebraska, where he dropped the ice in a government storage unit. That's 4,000 miles over a period of two weeks. This was hardly the only case.

The news media chronicled the stories of these truckers. A truck full of ice was sent from Dubuque, Iowa, to Meridian, Mississippi, then to Barksdale Base in Louisiana, then to Columbia, South Carolina, and finally to Cumberland, Maryland, where it waited for six days before being sent to

Bettendorf, Iowa, where the ice was unloaded. Another truck was sent from Wisconsin to Missouri to Selma to Memphis, Tennessee, before finally dropping off the ice in a storage unit. There were 4,000 drivers enlisted in this incredible charade. No one knows for sure how much ice ever got through or how much if any good it did.

In one of the first incidents reported of what was to be two weeks of catastrophe, a group of volunteer firefighters from Houston came to New Orleans wanting to help. They were told to wait. They waited 48 hours and were ordered to go back. A group of doctors from Maryland tried to get in but FEMA sent them on to the Red Cross, which said it could do nothing without the approval of federal health officials.

After the New Orleans mayor made a call for firefighters to come help, 1,000 volunteers were sent to Atlanta, where they were put in a conference room at the Sheraton Hotel and subjected to seminars on sexual harassment and other bureaucratic matters. They were then told that their job would be to distribute flyers with a message on it: call 1-800-621-FEMA. Many or even most of these well-trained people left town. Those who stuck it out and headed for Louisiana were aghast that their first assignment was not to fight fires, which had been raging for a week, but to escort President Bush on his TV-laden tour of the area. All the photos can be seen on WhiteHouse.gov.

In fact, FEMA refused offers of help of all sorts, mainly because of issues of control. FEMA declined help from Amtrak in evacuating people from New Orleans. The Chicago municipal government wanted to send volunteers from the fire department, police department, and hospitals all over town, but FEMA said no. The same happened to New Mexico, whose governor volunteered equipment and personnel.

FEMA prevented Wal-Mart from delivering three tank trucks full of water, and the Coast Guard from delivering diesel fuel. As for the National Guard, for days it would not allow reporters into the Superdome where tens of thousands were trapped. People were hungry and thirsty, but the National Guard would not allow the Red Cross to deliver any food. Here is the astounding statement from the spokesperson of the Red Cross: "The Homeland Security Department has requested and continues to request that the American Red Cross not come back into New Orleans. . . . Right now access is controlled by the National Guard and local authorities. . . . We cannot get into New Orleans against their orders."

The Salvation Army attempted to rescue two of its own officers trapped in a building and on dialysis. They rented three boats for a rescue. But they were not allowed through, though to be fair the Salvation Army didn't specifically name the government as at fault, but it did point out that all private efforts were running into similar kinds of obstacles, so the message was clear. Meanwhile, the USS *Bataan*, a floating hospital for 600 patients, that happened to be at sea and rode out the storm, was still sitting empty by the third day, not permitted to do its job.

An astounding case of ineptness comes to us from the case of three Duke University students who drove to New Orleans to help but were turned away by the National Guard. They had seen the news and knew that they could help, and wondered why they should be pushed around by bureaucrats. Being college sophomores, they took a risk. They forged press credentials, with fake IDs and shirts and the works. They went back and adopted a haughty tone. The National Guard waved them through immediately. Then the students drove to the Convention Center where they found thousands of sick, hungry, thirsty, and dying people in desperate need. They found a man who had welts all over his body. He was in a tree covered with fire ants as the flood waters rose, and there he stayed being bitten repeatedly for up to 24 hours.

The boys picked him up along with three others and drove them to a Baton Rouge hospital. They made another trip there and back with more people before they began to become frightened of what the government might do to them. On one return trip, they observed 150 empty buses driving the other way—and they have a video to prove it.

One can only express astonishment at how the government treated the tens of thousands of people it had herded like cattle into large public spaces. For reasons still unclear, the government couldn't seem to get its act together on transporting them out even as the people themselves were forbidden to leave. Once the central planners decided to move all these people from the Superdome to the Astrodome, no means of transport arrived, even as aerial photos showed miles and miles of public buses available.[11]

Indeed, the first bus to reach Houston was not driven or approved by the government. It was commandeered by 20-year-old Jabbar Gibson, who drove it from the floods and picked up as many people as he could and drove all the way to Houston, a 13-hour drive! He beat the government's system by a day. Meanwhile, the tens of thousands of people who had been shoved into the Superdome on Sunday, before the floods came, were still suffering in that massive calamity by Friday and Saturday.

Perhaps the most astounding case of incompetence has received the least attention. It relates to a 500-boat flotilla stretching over 5 miles that left for New Orleans from Lafayette. It involved 1,000 people who had hoped to rescue hospital patients and take them to safety. It consisted of private boaters, fishermen, hunters, and others who had spent their entire lives navigating Louisiana waterways. Once this caravan arrived, they were turned away by the Department of Wildlife and Fisheries, which was now being run by FEMA. All 500 boats were turned away and ordered out.

Now keep in mind that this was three days after the hurricane hit. There were hundreds of people inside the Charity Hospital in New Orleans alone.

They had no supplies and only three had been rescued. At this very time, the head of the FEMA-ized Wildlife and Fisheries Department announced to the world on television that it needed no help from anyone and that it had all matters under control.

Some White House staff members put together a DVD of the evening news coverage for Mr. Bush to watch on Air Force One, which was the only way they could get him to understand the depth of the crisis. The purpose of the action was not so much to help people, of course, but rather to stop the meltdown of his reputation. In fact, by the time he actually arrived in Louisiana, food and medicine deliveries, such as they were, were halted on orders of the White House, to make room for the presidential caravan.

In one particularly interesting detail, Katrina triggered the first use of the Department of Homeland Security's great accomplishment since it was created after 9-11: the National Response Plan, a 426-page central plan for dealing with a crisis on the level of the post-Katrina floods. Here is how the government describes it:

> The National Response Plan establishes a comprehensive all-hazards approach to enhance the ability of the United States to manage domestic incidents. The plan incorporates best practices and procedures from incident management disciplines—homeland security, emergency management, law enforcement, firefighting, public works, public health, responder and recovery worker health and safety, emergency medical services, and the private sector—and integrates them into a unified structure. It forms the basis of how the federal government coordinates with state, local, and tribal governments and the private sector during incidents.

What happened to the National Response Plan after the floods? It remained what it always had been: a colorful PDF download, a thick book on the management shelves, an item in the Government Printing Office catalog, bird-cage liner, and many other things. One thing it was not was a National Response Plan that did all those glorious things listed above. As with all such plans from time immemorial, it was a dead letter.

As for the National Guard, when it arrived it did what a military does best: harass the residents. Working with the police, it began to enforce an order for everyone to evacuate. The *New York Times* summarized the order as: "No civilians in New Orleans will be allowed to carry pistols, shotguns, or other firearms of any kind."

The National Guard allowed themselves to be videotaped going from house to house, and mansion to mansion, knocking down doors, searching for weapons, handcuffing owners and humiliating them. They called these people "holdouts," a phrase right out of Baghdad. One National Guardsman was asked whether he would shoot residents if they resisted. Yes, he would, he said. He added, "It's surreal. You never expect to do this in your own country."

We have provided a look at some of the terrible failures by the government—not only failing to do what it claimed it would do but actively working to prevent others from helping. The cost to human life and prosperity is incalculable. But, one might say, at least the government is generous now in preparing to spend perhaps $250 billion to clean up and reconstruct what was destroyed.

But who will get this money and where will it go? Cynics could not be more correct: the first company to receive the money was our old friend Kellogg Brown and Root (KBR), a current client of Mr. Bush's former campaign manager and former head of FEMA. KBR is a subsidiary of Halliburton, the company formerly headed by Dick Cheney. Another winner is Bechtel, whose former head is now in charge of Bush's Overseas Private Investment Corp. The top rebuilding priority: repairing government military bases in Louisiana and Mississippi. Those who work for one of these companies will do very well by this aid. As for the victims, they can expect little from the monstrosity that taxes and controls them relentlessly on the pretext that it will protect them and care for them when no one else will.

Fortunately for the people who lived in flooded areas, they did not face the crisis alone. The private commercial sector, along with thousands of religious charities, was there to help. Indeed, John Tierney of the *New York Times* was one of the few mainstream journalists to notice that Wal-Mart improved its image after Katrina. Its stores in the disaster-stricken areas still carried generators. Wal-Mart trucks rode into areas immediately following the hurricane and gave away chain saws, boots, sheets, and clothes for shelters, plus water and ice. It alone had prepared for emergency with its own emergency operations center.

Westley[12] noted that Wal-Mart gave $20 million in cash donations, 1,500 truckloads of free merchandise, food for 100,000 meals, and the promise of a job for every one of its displaced workers. After comparing FEMA's failures with Wal-Mart's successes, he concluded that the former ought to be abolished. Tierney, in contrast, drew the wrong conclusion. He said that the Wal-Mart CEO "is the kind of leader we need to oversee the tens or hundreds of billions that Washington will be spending on the Gulf Coast. [President] Scott could insist on low everyday prices while still leaving the area as well prepared for the next disaster as Wal-Mart was for Katrina."

In fact, if Lee Scott were given a government job, it would only be a matter of time before he became just another Michael Brown, the disgraced former FEMA head. This isn't a matter of character. It is a matter of the maze in which people find themselves, with market-generated exit signs, or none at all, thanks to the government.

As Block and Thornton[13] and many others have shown, it was not the storm as such that did the damage, but the failure of the government at all levels, particularly with regard to levees. Combined with the levees-only river management strategy of the Army Corps of Engineers, the floods were a disaster waiting to happen. Just imagine if the entire Big Easy were private, like homes or cars. Insurance companies would have taken a huge role in risk assessment, not only charging more for higher risk but insisting on management strategies that reduce risk and rewarding those who adopt those strategies with better premiums. This works on the same principles as homeowners' insurance, which combines rules and incentives to reduce the likelihood of losses.

Government insurance, however, makes us less cautious and more willing to take risks. It prices coverage from losses far too low and creates an environment where disasters like flooding are waiting to happen. With programs like subsidized flood insurance, government is like a bad mother who pays her children to run around with scissors. Government ownership is even worse because there are no signaling systems in operation at all. It was also the government that created a false sense of security for people in New Orleans, who were led to believe that the levees would hold and pumps would work. And when the floods finally did come, they were told the government would be there to manage the crisis.

But the government cannot manage crises, as the response to Katrina demonstrated. The local government fled. The state government dithered. And the federal government actively worked to prevent good things from happening. All the acts of private heroism that took place after Katrina occurred were despite government and not because of it.

And yet what lessons does the political culture want us to take from this? It is the same lessons we are instructed to learn after NASA spends and spends and still can't seem to make a reliable space shuttle. We are told that NASA needs more money. The public schools absorb many times more—thousands times more—in resources than private schools, and still can't perform. So we are told that they need more money.

The federal government spends trillions to "protect" the country and can't fend off a handful of malcontents with an agenda. And so we are told that the government needs to start several new wars and erect a massive new bureaucracy and put sections of the country under martial law at the slightest sign of trouble. And, spend more money.

So too, Congress can allocate a trillion dollars to fix every levee, fully preventing the last catastrophe, but not the next one. The real problem is the same in all these cases, not insufficient resources but public ownership and management.

Public ownership has encouraged people to adopt a negligent attitude toward even such obvious risks. But private developers and owners, in contrast, demand to know every possible scenario as a way to protect their property. But public owners have no real stake in the outcome and lack the economic capacity to calibrate resource allocation to risk assessment. In other words, the government manages without responsibility or competence. Actually, it was Mr. Bush who said one of the most sensible things, on September 2, 2005, "If you want to help, if you're listening to this broadcast, contribute cash to the Salvation Army and the Red Cross. . . . They're on the front lines providing help to the people who need help."

But it was two weeks later when his more basic instincts kicked in and he delivered a very different message, one that is deeply alarming. He said: "It is now clear that a challenge on this scale requires greater federal authority and a broader role for the armed forces—the institution of our government most capable of massive logistical operations on a moment's notice."

It is interesting that his beloved military was not there at a moment's notice. It now cites its own failures as the great excuse to expand its powers. So the government will spend the next several years preparing for another Katrina that may never come, just as it spent the last several years preparing for another 9-11 that may never come. The next crisis will be something completely unexpected, and once again it will fail. But we will be left with a government with some very bad habits, among which are declarations of martial law, mandatory evacuations, gun and price controls, and other totalitarian policies, and its attachment to military means, we get what can only be described as the continuation of the fascist track: the militarization of the country under its armed forces.

So far as we know, this passing remark by Mr. Bush has provoked no commentary in the national press. Commentators in the organized conservative movement have displayed an appalling deference to the administration's priorities, with *National Review* consistently arguing for more spending and militarization, Rush Limbaugh calling for price gougers to be strung up, and even some supposed free market friends calling for billions to rebuild New Orleans as a way of showing terrorists that we won't let the weather get in the way of progress.

Conservatives have been especially bad at tolerating egregious uses of the military. We need to reflect on what it means to have the military takeover in the event of crisis. What kind of ideology promotes such things, and looks the other way when it happens? This is clear, and it serves as a reminder that not all threats to freedom come from the left. A clue comes from the neo-Nazi novel called *The Turner Diaries*, sometimes cited as the motivating force for the bombing of the Oklahoma federal building, which ends in what the author regards as the utopian political system. After a world war that exterminates all non-Whites, a military regime takes over the United States and centrally plans the economy under martial law. All food and water are distributed on military trucks, all production takes place on a planned basis, and the merchant class is required to obey or be shot. The state also places flower boxes under every window.

The author describes this race-based national socialism as if it were a system with an inherent appeal to the reader, and perhaps there are people economically ignorant enough and full of enough loathing for humanity and freedom to regard it as attractive. In our own times there are people waiting in the wings who long for power and who are drawn to the ideal of a militarized society and a centrally managed economy. Some call themselves conservatives, and they are as much a threat to civilization as those with the same ideas who call themselves liberals.

Despite the foregoing, there is still room for some optimism. The government cannot actually do what it promises, and there is a way in which we can only be thankful for that. It cannot succeed in managing a central plan. Its plans will always fail. The government tries to use its failures as an excuse for more power, but with every failure comes a substantial degree of public humiliation for the public sector, and that humiliation can provide a basis for the undoing of government authority.

Some people say that a loss of government authority will mean the breakdown of civilization. Actually it will create the preconditions for the reestablishment of civilization, and in a state of freedom that can happen very quickly. The aftermath of Katrina illustrated in a million individual acts of charity and enterprise that people can manage their affairs, even amid the chaos. The calamity following Katrina was an egregious display, one that gave the federal government a black eye. The Democrats will continue to use this to harm the Republicans, which is fine, but it is not just Mr. Bush that is suffering, but the whole apparatus of central planning by decree from above. A government that cannot manage a crisis should not be trusted to manage anything at all. Thanks to Katrina and its dreadful aftermath, it is fair to say that the age of not trusting government has returned, and with a vengeance.

It took decades for the rot to give away underneath the Soviet apparatus of central planning. But eventually the implausibility of the entire project was no longer possible to deny. It gave way under an intellectual reaction against the whole of socialism. We are seeing something like that take place today, as government fails in Iraq and New Orleans and in every place around the country and the world where it causes problems and creates no solutions. The age of confident central planning is behind us. Right now, the state is just trying to keep its head above water. If freedom is to have a future, the time will come when it will sink to an ignoble end, and we will wonder how we ever believed in this myth called government crisis management.

The advocates of freedom and the partisans of private enterprise will be there with the intellectual equivalent of flotillas, barges, buses, helicopters, and the whole apparatus necessary to rescue liberty from every attempt to kill it. And when our City on a Hill comes to be, it will be privately built to withstand any flood.

NOTES

1. It is only praxeological reasoning which can establish apodictically that a system of private property rights and free enterprise is compatible with an economy based on rational principles, while one based on central control and central planning is not. See on this Hoppe, Hans-Hermann. 1992. "On Praxeology and the Praxeological Foundation of Epistemology and Ethics," Herbener, J., ed., *The Meaning of Ludwig von Mises*, Boston: Dordrecht; Mises, Ludwig von. 1998. *Human Action: The Scholar's Edition,* Auburn, Ala.: The Mises Institute; Rothbard, Murray N. 1951. "Praxeology: Reply to Mr. Schuller." *American Economic Review*, December.

2. In the felicitous terminology of Nozick, Robert. 1974. *Anarchy, State and Utopia,* New York: Basic Books, p. 163) all "capitalist acts between consenting adults" would also be legalized.

3. The best overall statements of libertarianism include Rothbard, Murray N. 1978. *For a New Liberty*, New York: Macmillan; Rothbard, Murray N. 1982. *The Ethics of Liberty*, Humanities Press, Atlantic Highlands, N.J.; Hoppe, Hans-Hermann. 1993. *The Economics and Ethics of Private Property: Studies in Political Economy and Philosophy*, Boston: Kluwer; Hoppe, Hans-Hermann. 2001. *Democracy - The God That Failed: The Economics and Politics of Monarchy, Democracy, and Natural Order*, Rutgers University, NJ: Transaction Publishers and Rockwell Llewellyn H. Jr. 2003. *Speaking of Liberty.* Auburn, AL: Mises Institute. For libertarian analyses of foreign policy see Higgs, Robert. 1987. *Crisis and Leviathan: Critical Episodes in the growth of American Government*, New York: Oxford University Press and Denson, John V., ed. 1998. *The Costs of War.* New Brunswick, NJ: Transaction Publishers.

4. Mises, Ludwig von. 1981 [1969]. *Socialism*, Indianapolis: LibertyPress/LibertyClassics.

5. Hayek, F.A. 1945. "The Use of Knowledge in Society." *American Economic Review.* Vol. 35, No. 4: 519–530; Stigler, George J. 1961. "The Economics of Information," *Journal of Political Economy* 69, pp. 213–225.

6. The "women and children first" policy was followed at the *Titanic* disaster. See on this http://www.anesi.com/titanic.htm; accessed on 8/23/06.

7. The Public Choice School is associated with this position. See Buchanan, James M., and Gordon Tullock. 1962. *The Calculus of Consent: Logical Foundations of Constitutional Democracy*, Ann Arbor: University of Michigan. But also see these Austrian critics of that perspective: Rothbard, Murray. 1997. "Buchanan and Tullock's 'The Calculus of Consent,'" *The Logic of Action II*, UK: Edward

Elgar Publishing Limited, pp. 269–274; DiLorenzo, Tom and Walter Block. 2001. "Constitutional Economics and the Calculus of Consent," *The Journal of Libertarian Studies*, Vol. 15, No. 3, Summer, pp. 37–56.

8. Anderson, William L. 2000. "The Dangers of Smart Growth. *The Free Market*, May 28, Vol. 18, Number 2; http://www.mises.org/fullstory.aspx?control=434&id=75.

9. Murphy, Robert. 2006. "The Gas-Line Quagmire in Iraq" February 6. http://www.mises.org/story/2026; Westley, Christopher. 2005. "The Right to Set Your Own Price." October 21; http://www.mises.org/story/1942.

10. Rothbard, Murray N. 1971B. "Nixonite Socialism," *The Libertarian Forum*, Vol. 3, No. 1 January, pp. 1–2; http://www.mises.org/story/1875.

11. Here are some of the pictures of the submerged school buses, located near the New Orleans "Sewer" Dome: http://billhobbs.com/hobbsonline/007188.html.

12. Westley, Christopher. 2005B. "FEMA Should Be Shut Down." September 12; http://www.mises.org/story/1908.

13. Block, Walter. 2006. "Katrina: Private Enterprise, the Dead Hand of the Past, and Weather Socialism; An Analysis in Economic Geography." *Ethics, Place and Environment: A Journal of Philosophy & Geography*; Vol. 9, No. 2, June, pp. 231–241; Thornton, Mark. 1999. "The Government's Great Flood," September, Vol. 17, No. 9; http://www.mises.org/freemarket_detail.asp?control=8&sortorder=articledate.

Chapter 14

Weather Socialism

INTRODUCTION

The Gulf Coast area of the United States has been subject to a series of hurricanes, most recently Katrina, which has killed an estimated several thousand people as of this writing. The storm missed New Orleans by a whisker and yet was responsible for mass devastation and flooding in this city. The Big Uneasy lies several feet below sea level and is surrounded on three sides by the Mississippi River (which wends all around it) and Lake Pontchartrain.

Even apart from such calamities, storms are a major inconvenience. They interfere with plans, meetings, seminars, writing, reading, classes, ordinary living, and so on. Storm warnings present denizens of New Orleans with a stark choice: evacuate at great inconvenience and run the risk that the oncoming bad weather will veer elsewhere rendering such flight unnecessary, or stay put and brazen it out, and have to deal with flooding, lack of electricity, no air conditioning in 90-degree-plus temperatures, to say nothing of actual loss of life and/or limb, either due to the storm itself, or to the aftermath, including looting, flooding, and loss of law and order.[1]

The present chapter attempts to provide an economic and geographical analysis of this tragedy. Section II is devoted to a discussion of the role free markets and reliance on private property rights could have but were prevented from playing in the Katrina debacle. In section III we discuss a theory of economic geography, focusing on the role that past decisions play in present settlements in geographical space. Section IV is given over to some speculations as to the role private enterprise can play in ameliorating weather conditions. We conclude in section V.

PRIVATE ENTERPRISE

First of all, the levees that were breached by the hurricane were built, owned and operated by *government*,[2] specifically by the Army Corps of Engineers.[3] The levees could have been erected to a greater height. They could have been stronger than they were. The drainage system could have operated more effectively. Here, the New Orleans Sewerage and Water Board[4] was at fault. It consists of three main operating systems: sewerage, water, and drainage. Had they been run more effectively, a lot of the inconvenience, fright, and even loss of life undergone in this city could have been avoided.

Public-sector creation of these facilities may have fooled many into thinking they were safer than they were, in actuality. People were in effect subsidized by these governmental expenditures, and encouraged to settle in the Big Easy. Without this particular bit of public-sector mismanagement, New Orleans would likely have been settled less intensively. (On the other hand, at one time this city was the largest in the South; statist negligence of a different kind —graft, corruption, over regulation—is responsible for it having a smaller population than otherwise.)

We should not be appalled by these failures. After all, it is only human to err. Were these levee facilities put under the control of private enterprise, there is no guarantee of zero human suffering in the aftermath of Katrina. No, what must enrage us is not any one mistake, or even a litany of them, but rather the fact that there is no automatic feedback mechanism that penalizes failure,[5] and rewards success—the essence of the market system of private enterprise. Will the Army Corps of Engineers or the New Orleans Sewerage and Water Board suffer any financial reverses as a result of the failure of their installations to prevent the horrendous conditions now being suffered by New Orleanians? Will they go bankrupt and no longer reduce the well-being of society with their inefficiency as a result of loss of profits? To ask these questions is to answer them.

One crucial step forward then, would be the privatization of such enterprises as part of the rebuilding process.[6] Perhaps a stock company could be formed; it is likely that the largest hotels, restaurants, universities, hospitals, and other such ventures would have an incentive to become owners of such an initiative. Right now, the levees are run by the very same type of folk responsible for the post office and the motor vehicle bureau.[7] We need to take no position on whether levees are a good or bad thing; only that if they are to be built, this should be done by an economic entity that can lose funding, and thus put its very existence at risk, if it errs, and creates havoc. This can only apply to the market, not the state.[8]

This is neither the time nor place to examine in detail the case for private ownership of bodies of water such as the Mississippi River, Lake

Pontchartrain, and, indeed, all oceans, rivers, seas, and lakes.[9] But the same principles apply here as they do to land. Suffice it to say that this is a question that should be explored, for it is no accident that where there is private property there is safety and responsibility, and where there is not, there is not.

Second, numerous roads, highways, and bridges were washed out, collapsed, or swept away. This makes it far more difficult for rescuers to get to the beleaguered city, and for refugees to leave. The enterprise that built, operated, and maintained these failing facilities was government.[10] Various oil drilling rigs also failed, and were swept away in the maelstrom. It of course cannot be denied that these were all private enterprises. One of them even collided with a bridge, greatly damaging it. However, there is a significant difference between the two types of events. The market test of profit and loss applies only to the latter, not to the former. Those oil companies that built their platforms more strongly will tend to grab market share from those that did not. No such regimen operates in the governmental sector. Imagine if the oil drilling rigs were all built by the state. They would have undoubtedly created far more damage.

THE DEAD HAND OF THE PAST

It is by no means clear that there should even *be* a city in the territory now occupied by New Orleans.

Ideally, under a regime of economic freedom, what determines whether a geographical area should be settled at all, and if so how intensively? It depends upon whether or not, in the eyes of the human economic actors involved, the subjective costs outweigh the benefits. The reason no one lives in the north or south poles, and that population density in Siberia, Northern Canada, and the desert areas of Nevada is very low, is that the disadvantages are vastly greater than the advantages in those places.

However, if government subsidizes building in areas in which people on their own would not choose to locate, then the populace will not allocate itself geographically in a rational manner. Similarly, the government declares drought-stricken farmland an emergency area, and heavily subsidizes agriculture in such locales,[11] there is also misallocation of settlement in this regard.

The Federal Emergency Management Agency (FEMA)[12] was created as part of the U.S. Department of Homeland Security only on March 1, 2003, but the federal government has been doling out gobs of money to inhabitants of areas struck by tornados, storms, snow, and other inclement weather for years. Such declarations number in the dozens for 2005 alone.[13] Southern Louisiana, Mississippi, and Alabama have already been declared federal disaster areas. Tons of money will pour into these political jurisdictions.

Thus, locational decisions are and will continue to be rendered less rational than otherwise, if people had to pay the full costs of their geographical settlement decisions.

It may well be that with the advantage of hindsight, the Big Easy is like several of these other places: not too cold, or drought-stricken, but too low, below sea level, and thus too much in danger of being flooded.

States Mises in this regard:

> Suppose that, making use of our entire store of technological skill and our present-day knowledge of geography, we were to undertake to resettle the earth's surface in such a way that we should afterwards be in a position to take maximum advantage of the natural distribution of raw materials. And suppose further that for this purpose the entire capital wealth of the present were at our disposal in a form that would allow us to invest it in whatever way was regarded as the most suitable for the end in view.
>
> In such a case the world would certainly take on an appearance that would be very considerably different from the one it now presents. Many areas would be less densely populated. . . . The great trade routes would follow other courses.[14]

At first glance, this does not support a New Orleans with anything like its close to one-half million population at its present location. Yes, this city is situated at the mouth of a great river, and offers a world-class port to international shipping. On the other hand as recent events have so tragically demonstrated, these benefits may be more than offset by the fact that it lies below sea level.

Does this mean that New Orleans would be doomed under a free enterprise system? This is quite possibly the case, if we could do everything all over again, and start off de novo, at the present time. But not necessarily, given that vast investment has already been made in streets, buildings, pipes, wires, parks, and so on. Even though, perhaps, if we knew then what we know now, no city would have been erected south of Lake Pontchartrain, it does not logically follow that it would not be rebuilt by private enterprise at present, under realistic assumptions. Given that New Orleans is now located where it is, it is entirely possible that it is economical for there to remain a large human settlement in that area. What cannot be denied is that when government enters the picture, economic calculation of this sort becomes impossible.

Mises (2003, p. 234) continues his analysis:

> With regard to choice of location . . . new plants appear most efficient in the light of the existing situation. But . . . consideration for capital goods produced in the past under certain circumstances makes the technologically best . . . (location) . . . appear uneconomical. History and the past have their say. An

> economic calculation that did not take them into account would be deficient. We are not only of today; we are heirs of the past as well. Our capital wealth is handed down from the past, and this fact has its consequences. . . . (S)trict rationality . . . induces the entrepreneur to continue production in a disadvantageous location.

That is, New Orleans might well be a "disadvantageous location" based on the assumption that we can with hindsight rearrange all previous locational decisions. But, we cannot do any such thing. Rather, capital (buildings, roads, pipelines, etc.) are bequeathed to us *at a certain location.* As it happens, lots of valuable capital is located in New Orleans. This fact would incline us to reinvest in that locale, storms be damned. But only the free market can make such a decision on a rational basis. When government muddies the waters, this cannot take place.

The best way then to rationally determine whether or not the Big Easy should be saved, is to leave this decision entirely to free enterprise—to capitalist entrepreneurs, who, alone, can rationally make such determinations. As the Austrian side of the Socialist Calculation Debate[15] has demonstrated, only with market prices can this be done. Moreover, private owners make such decisions with their own money, or funds entrusted to them; if they err, they alone suffer. They do not bring the rest of us down along with them.

WEATHER SOCIALISM

But there is a third element we cannot ignore: weather socialism.

According to an old adage, critics of government can properly blame this institution for many things, but bad weather is not among them. Wrong, wrong. At the risk of sounding out of step with the mainstream, the state is responsible not only for hurricanes, but for tornados, storms, typhoon, tsunami, excessive heat, excessive cold, too much rain, too little rain, floods, droughts, desertification, tempests, squalls, gales, rainstorms, snowstorms, thunderstorms, blizzards, downpours, cyclones, whirlwinds, twisters, monsoons, torrential rains, cloudbursts, showers, and so on. You name any kind of bad weather conditions, and the government is to blame.

Why, pray tell? Because the state at all levels commandeers almost 50 percent of the GDP in taxes, and its regulations account for a significant additional amount of wealth not created.[16] If the voracious government left all or even most of the property created by its rightful owners—those who created it in the first place with their own efforts—the weather problem could undoubtedly be better addressed by private enterprise.

Let us consider the following argument, critical of the above claim:[17]

> Individuals would not be better able to protect themselves against natural hazards if they lost fewer resources to government taxation. For the bottom half of income earners today pay almost no taxes, which means that a reduced take by the government could affect them only indirectly, as lower prices flowing from increased capital investment. Because so many of its residents are poor, New Orleans was likely a net beneficiary of Federal money. So cutting taxes would cost New Orleans money, with very little chance that the funds, now restored to private hands, would be invested in New Orleans. Most residents of New Orleans would be worse off, at least in the short run, if redistribution programs were curtailed. Tax relief grows less and less attractive as taxes grow more and more progressive, while at the same time increasingly progressive taxation increases the number of places with a stake in maintaining the inertia of an inefficient settlement system.

Let me say the following in response. First, it is only with regard to income taxation that the poor pay very little; they pay a disproportionate share of other taxes, however. Since food, clothing, certain types of entertainment, and so on, play a disproportionately high role in their budgets, government levies on these items impact them heavily. Second, while income taxes are not directed at the poor, there is such a thing as tax shifting; it is by no means clear that the poverty-stricken escape to the extent that would otherwise be the case were this consideration not taken in to account.

Third, while, indeed, a rose is a rose is a rose, it is not the case that a dollar is a dollar is a dollar. That is, a dollar emanating from the government, via the good graces of the ilk of FEMA, is not at all equivalent to a dollar in the private sector. All too many FEMA dollars find their ways into the pocketbooks of politicians and their cronies. Then, too, there is outright waste as when tens of thousands of mobile homes, renting for an astounding $3,000 per month and more, sit rusting far away from New Orleans in the fields of Arkansas of all places. Even were these trailers delivered to their supposed relatively poor New Orleans recipients in a timely manner, which they certainly were not, this *still* represents vast waste. For it is unlikely that the recipients would have spent the money in precisely this manner. Which would be more valuable to you: $10,000 cash money in your pocket, or a basket of goods costing just that much money on the market, consisting of, oh, say, a violin, a bicycle, a year's supply of ketchup, 100 meals at McDonalds, and so on? It is only in the unlikely case that the recipient would have spent the money in *precisely* this manner that there is no loss. But that would be a rare case indeed. However, the same logic applies to giving people trailers, instead of the money they cost, even on the assumption of no vast price inflation suffered by inept and corrupt government purchasers.

Fourth, this objection reckons in the absence of the insight that in the market, the rich protect the poor as a sort of byproduct or externality. For example, the internal "streets" in shopping malls are vastly safer than the outdoor streets provided by the government. The poor, along with everyone else, are of course welcomed to shopping malls. There, they enjoy the enhanced safety provided by relatively wealthy shopping mall owners. How would this work in the New Orleans case? If associations of private businesses (i.e., wealthy people) were placed in charge of the levees, instead of the Army Corp of Engineers, it is likely in the extreme that the poor of the Big Easy would have been far better protected from Katrina than in fact they were.[18]

For what wondrous tasks does the government waste trillions of our earnings? Let me count a few of the ways. It subsidizes farmers who ought to be allowed to go bankrupt when they cannot earn an honest profit in this industry. As the numbers of farmers have declined over the years, the number of bureaucrats in the Department of Agriculture has increased. This constitutes nothing but welfare for farmers, and agricultural mandarins. Speaking of welfare, this is but the tip of the iceberg. Our masters in Washington D.C. distribute our hard-earned money to recipients who parent children they cannot afford to feed, and to corporate welfare bums, mainly limousine liberals. Then there is the Department of Education (weren't the Republicans supposed to repeal this sore on the body politic?) that presides over a public school system that warehouses and mis-educates our children. Then there is our system of medical socialism that wastes yet other precious resources.[19] We don't have Hilary-Care, yet, but we are well on our way. Then, too, we must count government throwing our money at the Post Office, the Space program, ethanol, foreign "aid," unemployment insurance, the list goes on and on.

The drug war incarcerates thousands of innocent people—who could be creating additional wealth—at a cost exceeding tuition and room and board at some of our most prestigious universities. Last but certainly not least, speaking of war, the United States has been bullying its way around the world for decades, creating untold havoc. Katrina can't hold a candle to our armed forces in terms of killing innocent people. There are no truer words than that "War is the health of the State."[20]

Suppose that the "public sector" was not wasting untold riches. What has this got to do with improving weather conditions? Well, a lot of the money returned to the long-suffering taxpayers (and much of the additional wealth created by the ending of economic regulations) would be allocated in the usual directions: sailboats and pianos, and violin lessons and better food and more entertainment, and so on. But some of it would likely be invested in more research and development as to the causes and cures of unwelcome weather conditions. Is there any doubt that in 100, or 1,000, or 10,000 years—assuming the government does not blow us all up before then—we will no longer be plagued by uncooperative clouds? I don't say that if the state disappeared tomorrow the next day we would have clear weather (and rain to order from 2:00 am to 4:00 am), but surely the ending of the former would bring about the latter that much quicker.

How would this work? Would not the problem of "public goods" rend the market a "failure," as members of the so-called Chicago "free enterprise" school of economics would have it? Their argument[21] is that if I come up with a way to stop storms dead in their tracks, or, better yet, orchestrate matters such that they do not form in the first place, everyone else will "free ride" on my innovation. The other beneficiaries will simply refuse to pay me for this boon I confer on them, so I will not invest any money on this task in the first place. And neither will you. So the private enterprise system cannot handle such challenges.

This thesis is highly problematic.

First of all, this task need not be accomplished on a for-profit basis. Nonprofit organizations, too, are part of the private sector of the economy. Just looking at the charitable outpourings to New Orleanians from all corners of the country, we can see that there is no shortage of benevolence and goodwill for the victims of Katrina. I should single out for special mention in this regard that "evil" profit-maximizing large corporation that grinds down suppliers, immiserates its own workers due to its anti-union policies, bankrupts small grocers, and just all around exploits everything else it touches: Wal-Mart. This hated corporation contributed $1 million to the Salvation Army[22] for hurricane relief. More recently, Wal-Mart committed an additional $15 million for this purpose. As part of this commitment, Wal-Mart will "establish mini-Wal-Mart stores in areas impacted by the hurricane. Items such as clothing, diapers, baby wipes, food, formula, toothbrushes, bedding and water will be given out free of charge to those with a demonstrated need."[23]

The point is, if we the people had vastly more money at our disposal than we do now, thanks to government profligacy with our funds, we would be able to donate some of it to the not-for-profit sector to engage in research and development for weather control.

Second, the market has a way of internalizing the so-called externalities that supposedly prevent firms from providing storm-busting services.[24] Within limits, and depending upon technology, the purveyors of flood insurance would be able to turn the rain and wind on and off like a spigot, depending upon the locational densities of their clientele. For example, if in area A 90 percent of the landowners are members of Hurricane Busters, Inc., and in area B only 10 percent are, there is little doubt as to which will be better served by this particular firm. Then, too, there will be not only social pressure, but economic pressure, for large firms in any geographical area to sign up for such services. Those that do not (particularly those in states stretching from Texas to Florida, and most certainly in New Orleans) will tend to find their customer base disappearing.

As but one small instance of this phenomenon, companies with large parking lots have recently instituted reserved spaces for pregnant women and new mothers.[25] No government agency forced them to do any such thing. (Prediction: the state will soon do just that, so as to garner credit for this very human and profitable policy.) As this movement catches on, few will be able to resist. A similar situation is likely to arise with regard to protection from hurricanes. At the very least, if government would but get out of the way, it would clear the path for private enterprise to more quickly bring us the day when the Katrinas of the future will be obviated.

CONCLUSION

What is the libertarian position on the storm and its aftermath? No national guard or other representatives of the state should be brought in. They are in effect "murderers and thieves."[26] Instead, private police agencies, appointed by property owners, should deal with the looters. Further, no tax money should be poured into New Orleans. These are stolen funds, and should be returned to their rightful owners, the taxpayers of the nation. Of course, this applies, in spades, to those victimized by Katrina. But the refunds should be in the form of money to specific taxpayers, not generalized expenditures for rebuilding, which their proper owners may or may not favor. Instead, New Orleans should rely on private charity.[27]

Private enterprise should alone determine if the Big Easy is worth saving or not. Problems of "transactions costs" will be far easier to overcome than challenges presented by an inept and economically irrational government. Possibly a Donald Trump type might try to buy up all the buildings at a fraction of their previous value and save his new investment by building levees and water pumps. He wouldn't need to purchase 100 percent of all real estate. A lesser amount, say, 90 percent, might do, and he would only make his initial purchases subject to reaching this level. That is, he might first purchase options to buy.

NOTES

1. I have been a resident of New Orleans since fall 2001. Many of the considerations in this chapter emanate from that experience. I wish to thank a referee for several suggestions that greatly improved an earlier draft of this chapter. I alone of course am responsible for any remaining errors.

2. Block, Walter. 2004. "Me and hurricane Ivan." September http://www.lewrockwell.com/block/block44.html.

3. See Rockwell, Llewellyn H. 2005. "The State and the Flood." September 2. http://www.lewrockwell.com/rockwell/flood.html.

4. See http://www.google.ca/search?hl=en&q=New+Orleans+Water+Board&btnG=Google+Search&meta=; http://www.bgr.org/budgets/s&wb/systems_description.htm.

5. Hazlitt, Henry. 1979. *Economics in One Lesson.* New York: Arlington House Publishers.

6. If that indeed occurs; for more on this, see below, section III.

7. Block, Walter. 2001C. "Motor Vehicle Bureau." September http://www.lewrockwell.com/block/block8.html.

8. The Edsel was at one time emblematic of the failure of the marketplace. This is a misinterpretation of colossal dimensions. Rather, it was a failed line of the Ford Motor Company, which was very quickly jettisoned; http://www.libertyhaven.com/regulationandpropertyrights/tradeandinternationaleconomics/risefall.html). Imagine, though, if the government were responsible for such a disaster. Then, there would be little or no economic incentives to cut one's losses and end this product that was rejected by consumers. One might reasonably think of the U.S. Post Office as a governmental Edsel.

9. But see on this Whitehead, Roy and Walter Block. 2002. "Environmental Takings of Private Water Rights: The Case for Full Water Privatization," *Environmental Law Reporter*, October, pp. 11162–11176; Whitehead, Roy, Catherine Gould and Walter Block. 2004. "The Value of Private Water Rights: From a Legal and Economic Perspective," *Albany Law Environmental Outlook Journal*, Vol. 9, No. 2, pp. 315–343.

10. For a critique of road socialism, that is, government ownership and operation of streets and highways, see Block, Walter. 1996. "Road Socialism," *International Journal of Value-Based Management*, Vol. 9, pp. 195–207.

11. See http://www.fsa.usda.gov/pas/publications/facts/html/EMProcess04.htm.

12. See http://www.fema.gov/about/.

13. See http://www.fema.gov/news/disasters.fema?year=2005.

14. Mises, Ludwig von. 2003 [1933]. *Epistemological Problems in Economics*, New York: New York University, Third Edition, p. 231.

15. See http://www.google.com/u/Mises?hl=en&submit.x=0&submit.y=0&&q=socialist%20calculation.

16. DiLorenzo, Thomas J. 2004. *How Capitalism Saved America.* New York: Crown; DiLorenzo, Tom and James Bennett. 1999. *The Food and Drink Police: America's Nannies, Busy Bodies, and Petty Tyrants*, Rutgers, NJ: Transaction.

17. This rejoinder was offered to me by a referee; what appears in the text is an edited version of the objection.

18. Were any private enterprise half as inefficient as FEMA or the Army Corp of Engineers, there is little doubt they would have long disappeared from the scene in bankruptcy court. In the event, these two government agencies were given *additional* funds in the wake of their abject failures.

19. Friedman, Milton. 2001. "How to Cure Health Care." The Public Interest. Winter. http://www.thepublicinterest.com/archives/2001winter/article1.html.

20. Rothbard, Murray N. 1963. "War, Peace, and the State." *The Standard*, April, pp. 2–5; 15–16. http://www.mises.org/rothbard/warpeace.asp.

21. Friedman, Milton. 1962. *Capitalism and Freedom*, Chicago: University of Chicago Press, 30–32.

22. See www1.salvationarmy.org/usw/www_usw.nsf.

23. In contrast, I do not recommend the American Red Cross. I still have not forgiven them for being so politically correct that they accepted blood from homosexuals, and infected hundreds of people with AIDS (http://www.aegis.com/news/sc/1992/SC921205.html), many of them hemophiliacs (http://www.aegis.com/news/lt/1993/LT931113.html). Unhappily, from my own point of view, Wal-Mart, the evil firm that it is, sent another $1 million to the Red Cross.

24. Block, Walter. 1983B. "Public Goods and Externalities: The Case of Roads," The *Journal of Libertarian Studies: An Interdisciplinary Review*, Vol. VII, No. 1, Spring, pp. 1–34; Hoppe, Hans-Hermann. 1989. "Fallacies of the Public Goods Theory and the Production of Security," *The Journal of Libertarian Studies*, Vol. IX, No. 1, Winter, pp. 27–46; Hummel, Jeffrey. 1990. "National Goods vs. Public Goods: Defense, Disarmament and Free Riders," *The Review of Austrian Economics*, Vol. IV, pp. 88–122; Sechrest, Larry. 2003. "Private Provision of Public Goods: Theoretical issues and some examples from maritime history."

25. See http://www.mises.org/story/1898.

26. Spooner, Lysander (1870) 1966. *No Treason: The Constitution of No Authority and A Letter to Thomas F. Bayard*, Larkspur, Colorado: Rampart College http://www.lysanderspooner.org/notreason.htm.

27. See http://www.lewrockwell.com/akers/akers16.html.

Chapter 15

Katrina

GOVERNMENT AND THE KATRINA CRISIS

Katrina gave the Gulf region of the United States a lesson in how government fails. In the weeks following the hurricane, it became clear that most of the destruction was not the work of Mother Nature but those who are charged with preventing precisely such disasters as this.

To begin on a personal note, I am a professor of economics at Loyola University New Orleans. The university is currently shut down. Many students have gone elsewhere, and my colleagues and I have worked to find other places to live and work until the university reopens. I escaped the disaster in time, and was safely ensconced in Western Canada when the floods came. My heart goes out to those who were very much less fortunate.

Now to the lessons.

The levees that were breached by the hurricane were built, owned, and operated by *government*. Specifically, they were owned and maintained by the Army Corps of Engineers. The levees could have been erected to a greater height. They could have been stronger than they were. There might have been a better detection system. All of these factors are clear in hindsight. As to preparing for the future, that is not the government's specialty.

Similarly, the drainage system could have operated more effectively. Here, the New Orleans Sewerage and Water Board was at fault. It consists of three main operating systems: sewerage, water, and drainage. Had they been more effective, a lot of the inconvenience, fright, and even loss of life undergone in this city could have been avoided.

Then, too, these facilities may have fooled many people into thinking they were safer than they actually were. I know this applies to me. Thus, people were in effect subsidized, and encouraged to settle in the Big Easy. Without this particular bit of government mismanagement, New Orleans would likely have been settled less intensively, or possibly, residents might have had to bear the higher costs associated with keeping the sea at bay.

Actually, we will never know since there is no market feedback mechanism at work between the quality and functioning of the infrastructure and the costs associated with maintaining and insuring it against failure. Instead, the infrastructure on which life depended was entrusted to public officials, who are isolated agents, cut off from market oversight and accountability and impervious to change.

I am not appalled by these failures. After all, it is only human to err. Were these levee facilities put under the control of private enterprise, there is no guarantee of zero human suffering in the aftermath of Katrina. No, what enrages me is not any one mistake, or even a litany of them, but rather the fact that there is no automatic feedback mechanism that penalizes failure, and rewards success, the essence of the market system of private enterprise.

Will the New Orleans Sewerage and Water Board suffer any financial reverses as a result of the failure of their installations to prevent the horrendous conditions now being suffered by New Orleanians? To ask this question is to answer it. What about the ownership and maintenance of the levees by the Army Corps? It's not even debated.

Bush said in his speech that "city and parish officials in New Orleans, and state officials in Louisiana will have a large part in the engineering decisions to come. And the Army Corps of Engineers will work at their side to make the flood protection system stronger than it has ever been."

So there you have it.

In the same speech, he summed up the creed of government that every failure should be met with ever more power and money: "The system, at every level of government, was not well-coordinated, and was overwhelmed in the first few days. It is now clear that a challenge on this scale requires greater federal authority and a broader role for the armed forces—the institution of our government most capable of massive logistical operations on a moment's notice."

The idea of militarizing all crisis situations is particularly disturbing. In the days following the flood, the National Guard went house to house disarming people, which can only increase the likelihood of more looting. The troops broke down doors and handcuffed people solely for the crime of living in their own homes. This is what armed forces do.

These are steps backward. A crucial step forward would be the privatization of this enterprise as part of the rebuilding process (if that indeed occurs; for more on this, see below). Perhaps a stock company could be formed; I suspect that the largest hotels, restaurants, universities, hospitals, and other such ventures would have an incentive to become owners of such an enterprise.

Right now, the levees, pumps, and response teams are run by the very same types of folks responsible for the post office and the motor vehicle bureau. I take no position on whether levees are a good or bad thing; only that if they are to be built, this should be done by an economic entity that can lose funding, and thus put its very existence at risk, if it errs. This can only apply to the market, never the state.

This is not the place to examine in detail the case for private ownership of bodies of water such as the Mississippi River, Lake Pontchartrain, and, indeed, all oceans, rivers, seas, and lakes. But the same principles apply here as they do to land. Suffice it to say that this is a question that should be explored, for it is no accident that where there is private property there is safety and responsibility, and where there is not there is none.

Another government failure was overlooked in the days following the flood. Consider that many roads, highways, and bridges were washed out, collapsed, or swept away. This makes it far more difficult for rescuers to get to the beleaguered city, and for refugees to leave. You will never guess who built, operated, and maintained these facilities. Yes, it was government!

It of course cannot be denied that various oil drilling rigs also came unglued, and that these were all privately owned. One of them even collided with a bridge, greatly damaging it. However, there is a significant difference between the two types of events. The market test of profit and loss applies only to the latter, not to the former. Those oil companies that built their platforms more strongly will tend to grab market share from those that did not. No such regimen operates in the governmental sector. Imagine if the oil drilling rigs were all built by the state, then they would have undoubtedly created far more damage.

We must also consider questions of population density. Ideally, under a regime of economic freedom, what determines whether a geographical area should be settled at all, and if so how intensively? It depends upon whether or not, in the eyes of the human economic actors involved, the subjective costs outweigh the benefits. The reason no one lives in the north or south poles, and that population density in Siberia, Northern Canada, and the desert areas of Nevada is very low, is that the disadvantages are vastly greater than the advantages in those places.

However, if government subsidizes building in areas people on their own would not choose to locate, then the populace can no longer allocate itself geographically in a rational manner. Similarly, if the government declares drought-stricken farmlands an emergency area, and heavily subsidizes agriculture in such locales, there is also misallocation of settlement in this regard.

The Federal Emergency Management Agency, created in 1979, became part of the U.S. Department of Homeland Security only on March 1, 2003. The federal government has been doling out gobs of money to inhabitants of areas struck by tornados, storms, snow, and other inclement weather for years. Such declarations number in the dozens for 2005 alone. Southern Louisiana, Mississippi, and Alabama have already been declared federal disaster areas. Billions of dollars will pour into these political jurisdictions. Thus, locational decisions are and will continue to be rendered less rational than otherwise, if people had to pay the full costs of their geographical settlement decisions.

As it happens, lots of valuable capital is located in New Orleans. This fact would incline us to reinvest in that locale, storms be damned. But only private enterprise can make such a decision on a rational basis. When government muddies the waters, this cannot take place.

The best way then, to rationally determine whether or not the Big Easy should be saved, is to leave this decision entirely to free enterprise—to capitalist entrepreneurs, who, alone, can rationally make such determinations. As the Austrian side of the socialist calculation debate has demonstrated, only with market prices can this be done. Moreover, private owners make such decisions with their own money, or funds entrusted to them; if they err, they alone suffer. They do not bring the rest of us down along with them.

For anyone watching the rescue efforts after the flood, it was clear who and what was responding. It was not government, which only began getting its act together after the public relations disaster in the first week. Private charitable organizations—which are also part of the market economy—responded heroically and quickly and well. I'm particularly proud of the role the Mises Institute played in providing a means for a lone internet company in New Orleans to broadcast from the scene of disaster to the entire world.

Katrina led to a horrible calamity. The lesson it teaches us is one that all of history teaches: private enterprise offers hope for success whereas government offers only a guarantee of failure.

Chapter 16

Price Gouging

INTRODUCTION

The concept of "price gouging" during times of emergency, such as in the aftermath of Katrina (Block, 2006; Block and Rockwell, 2007; Rockwell and Block, 2010), often evokes quite an emotional and reactive response from people. The vast majority are outraged that stores and companies would increase their prices during a time of emergency. How could firms be so cruel as to take advantage of those needy people in such a desperate situation?

The problem is that people do not realize that, in times of emergency, the market price they knew before is no longer adequate. They view price gouging as opportunistic pricing on the part of the "evil" capitalists in order to enrich themselves at the expense of the consumer. Legislation has even been passed to prevent entrepreneurs from raising their rates (National Conference of State Legislatures, 2004). Government intervention is not the answer to this "problem." Upon exploring basic concepts of economics, one might glean a better perspective of the justification for raising prices during times of emergency, as well as what would happen if there were not laws preventing this very necessary practice.

In section 2 we address these "basic concepts" of economics. Section 3 is devoted to applying them to emergency situations, preeminently the dire plight of New Orleans and the Gulf coast after Katrina. We conclude in section 4.

BASIC CONCEPTS

Prices are to the economy as street signs are to the urban geography, as are the moving baton and the sheet music to the orchestra. They are signals;

they convey vital information. If street signs were taken down, and the conductor and the score removed from the musicians, chaos would ensue. Something close to a real-world example occurred in the early days of the Soviet Union, when prices were all but banned. Such pandemonium took place that even the died-in-the-wool Communists, the true believers, had to pull back from this particular abyss. They instituted the New Economic Plan, which brought back some semblance of market prices (Rothbard, 1971; Richman, 1981).

But only market-driven not any old prices can serve in this capacity. If they are to do their job, they cannot be arbitrary. Prices must reflect the importance people place on the items to which they refer and their relative scarcities, in the eyes of the economic actors. If street signs were randomly switched with each other, or if the notes and movements of the conductor became haphazard, they would do no good at all. It is exactly the same with prices. Arbitrary ones are no better than none at all. Here, again, the economy of the USSR serves as a baleful illumination. Owing to their Communist ideology, markets in raw materials and capital goods were legally forbidden. If it were not for black markets in the Second World, and their knowledge of prices in the Free World,[1] they would have been flying completely blind.[2] They would not have known, for example, whether to use wood, plastic, or metal for their rowboats. They would not have been able to determine whether the resources that went into the construction of a building were worth more, less, or the same amount as the final good itself.

It is only through the free interplay of supply and demand, which incorporates the comparative evaluations of all market participants, that meaningful prices can be constructed. It is no accident that to the degree that an economy has been free in this way, it has also been prosperous (Gwartney et al., 1996). Thus, paradoxically, introducing a bit of arteriosclerosis into the economy by enacting anti-"gouging" laws will reduce economic welfare, not increase it.

APPLICATIONS

During a natural disaster, the demand for batteries, gasoline, milk, candles, bottled water and ice among other things generally increases. Simultaneously, the supply of these goods often decreases due to damage from storms, forest fires, tornadoes, or earthquakes. Even without the "evil" capitalist getting explicitly involved,[3] the natural market price will automatically increase beyond its pre-disaster status. Most people do not take this into account when complaining about the "fairness" of post-disaster pricing. The increase in price as a result of shifts in supply and demand is the market's way to allocate scarce resources. When the market is prevented from doing so (most often[4] due to government regulation), scarcity in the form of shortages of vital resources is augmented.[5]

In an article discussing the "gouging" of gas prices Lieberman (2006) illustrates the following point:

> But high prices lead to solutions. They encourage producers to increase supplies and consumers to cut back on unnecessary driving. Over the longer term, they may even create opportunities for alternative fuels. This explains why oil and gas prices fluctuate and why no increase ever has proved permanent. Now, some have lost patience with this process and seek to short-circuit it. But that only leads to demand outstripping supply at the mandated price—which, in turns, leads to shortages, gas lines, and rationing. The temptation is great. Consumers paying $3 or more for a gallon of gas may like the idea of government stepping in and reducing it to $2. They are less likely to appreciate passing gas station after gas station with "closed" signs in the window.

When government intervenes in the market by passing legislation that prevents sellers from adjusting their price to what it should be after a disaster, the situation is worsened. Business owners are treated as criminals, while the "anti-gouging" enthusiast consumer gets to rob them of money that is rightfully theirs, according to the market. The government is a co-conspirator and the main proponent of this economic crime. Consumers who truly believe that the government is helping them in such situations are sadly mistaken. Anti-gouging legislation reduces incentives for stores to open or for outside sellers to come to the disaster area. Sellers do not want to enter a market in which there is little or no opportunity to earn a profit. It is price that rations the scarce resources and calls forth supply. All other things being equal, production increases as a result of increase in price. Implicit in this is that if prices are not allowed to increase due to government regulation, production will not increase. In times of emergency, when the demand for certain goods increases, this "helpful" government regulation is the very reason that the resources are not available to those who need them. Outside firms and entrepreneurs are naturally reluctant to enter such a tumultuous "market." This is perfectly rational and does not imply some absence of a conscience on the firm's part.

Anti-price "gouging" activists have manipulated people into thinking that there is some moral flaw surrounding the natural fluctuations in price that occur in post-disaster situations. By doing so, they are transforming basic economic theory into an issue of what is just and unjust. It should not be an issue at all. If the market were manipulated to make sure that everything was "fair" for its participants, there would be many conflicts within it. It would hardly be a market (perhaps a better term for this would be socialist regime). It would be flooded with businesses that are not needed. Firms that, under normal conditions, should not be in business would stay in operation, wasting precious resources that could be better used in some other endeavor. Government subsidies would be rampant. Simultaneously, firms that should prosper if general economic welfare were to be promoted would be forced out by taxes and price ceilings. That is "fair," is it not? High prices are not an issue. They are the market's way of allocating resources among all those who wish to use them. The media creates these "issues" for rating purposes. Politicians use anti-gouging legislations as a means to get re-elected or increase campaign donations.

Then Congressman (now Senator) Bernie Sanders believes that companies charging "high" prices need to be stopped for the sake of the consumer. Throughout his article, "Stop oil company price gouging," Sanders explores the supposed wretchedness of profits and big business. While the concept of Big Oil raising its prices in response to market forces and the concept of businesses doing so during times of emergency are not exactly the same, the argument against such organic practices follows the same logic. Businesses are taking advantage of consumers. Government must protect the consumers. Government must regulate, that is, punish, businesses. Sanders (2005) refers to businesses as "corporate behemoths," describes price adjustments by them as "company greed" and "market manipulation," and classifies the benefits enjoyed by Big Oil as "huge windfall profits" made "on the backs of consumers." This is the anti-gouging attitude.[6]

Sanders is clearly not a supporter of the free market. He demonstrates this through his condemnation of corporations and market prices as well as his urging of consumer aid in the form of government regulation. By implying that profits made by these companies are made at the consumers' expense, Sanders is blatantly ignoring the fact that the consumers engage in these transactions willingly. Clearly the price, although higher than they may have liked, was one at which they benefited from making the purchase; otherwise they would not have willingly done so. Through engaging in transactions, consumers are demonstrating that they prefer the good or service more than they prefer the amount of money they must pay for it. As long as neither party was coerced, both parties benefit from exchange. Congressman Sanders further illustrates his neglect for this concept by labeling the prices "anti-consumer." Very much to the contrary, legislation that prohibits price adjustments is the anti-consumer measure. Companies earn profits by satisfying the wants of customers. People want these products and are willing to pay for them. If these companies are not allowed to earn profits due to price controls, they may not develop the technology needed to increase supply to the maximum extent possible. They could even go bankrupt. How will the demand for oil and gas be satisfied if companies are no longer in operation or cannot afford to produce at capacity levels due to outdated technology? If Sanders wants to reduce the price of oil and gas, he should explore ways to reduce demand, such as urging his constituents to use these resources in lesser amounts.[7]

The main flaw in the anti-gouging argument is its lack of logic. Consumers and politicians want "fair" prices, but have not thought much past that point. They are so consumed with the actual price that they neglect to think about the implications of such regulations. In the face of disaster, people panic. It is human nature to do so. Prices lower than otherwise indicated by supply and demand conditions, combined with panic, result in hoarding. People flock to stores and buy as much as they can. Even with quantity limits, supplies run out before most people can get to them. This is where the government is supposed to step in and take care of its people, at least as far as the pundits are concerned; the state has stepped in, alright, but it is not helpful. A good example of this is federal and state response (or rather lack of same) to hurricane Katrina:

From Wal-Mart's satellite-based communications systems to FedEx's aircraft, U.S. business has in some cases managed to provide a swifter response to the initial impact of hurricane Katrina than the federal and state authorities.

But critics of the handling of the crisis by government agencies said they were slow in accepting offers of help.

Mary Landrieu, the Democratic U.S. senator from Louisiana, accused the Federal Emergency Management Agency of "dragging its feet" when Amtrak offered trains to evacuate victims.

"Offers of medicine, communications equipment and other desperately needed items continue to flow in, only to be ignored by the agency," she said.

Wal-Mart, the largest U.S. retailer, with over 100 stores and distribution centers in the affected areas, sent in trucks carrying water, food, and other emergency items in response to calls from local relief groups in the first days after the hurricane hit (Birchall et al., 2005; Block, 2006; Block and Rockwell, 2007).

While the concern is often that monopolies and oligopolies[8] might arise, free market prices would have the opposite effect. When the market is allowed to adjust without the "help" of the government, resources will flood into an area. It will now be profitable for out-of-towners to move supplies and services to places such as New Orleans, where disasters have struck. The now legal increase in market price will serve as a signal to the suppliers of goods that the demand has increased. In turn, these suppliers will produce more in order to accommodate the shift in demand. This is the efficient way to get resources to those who want and need them the most. Markets will be saturated with entrepreneurs and competition. The increase in supply will eventually bid down the price of these resources. Market forces will stamp out the "gouging" that solved the problem in the first place far more effectively than government intervention. Both firms and consumers will get valuable information from prices. They will ration the resources temporarily undergoing shortages (batteries, candles, milk, water, etc.) and call forth new supply. If the so-called price gouging is allowed by law, it will not last for long; it will vanish when its rationing and incentive effects on entrepreneurs are no longer needed.

In his article about Wal-Mart's entry into the generic drug market, Devon Herrick captured the impact of competition on prices. Wal-Mart approached the market with the plan to offer these drugs at a lower price than their competitors. As a result, firms such as Target and K-Mart are attempting to match or beat Wal-Mart's price. Concerning this fierce competition, Dr Robert Graboyes of the National Center for Policy Analysis is quoted as saying, "They've demonstrably driven down consumer prices nationally" (Herrick, 2007).

By enacting price controls, moreover, the government is restricting entry to the market. Firms are not legally able to realize free market rates for their goods and services. By reducing incentives for firms outside the area affected by the disaster, government is directly restricting competition. This perpetuates a misallocation of resources. Another affect of this government intervention in the market is the monetary loss to which firms in the price-controlled area are subject.

The anti-gouging perspective is one of business taking advantage of the poor consumer. The businessman is "evil" for wanting to make a profit or charge a market price for his resources. The consumer is helpless and has no choice other than to pay these astronomical prices. Keeping in mind the aforementioned economic theory of mutual gains from exchange, this concept is nothing short of irrational. By regulating and restricting business during times of emergency (or any time for that matter), the government is enhancing the economic loss suffered by the afflicted area. According to Malek, supporters of price ceilings believe that it is morally justified for the consumer to pay as little as possible and immoral for the firm to charge as much as possible. But to imply that a price is not fair (too high) is to imply that the consumer has some inherent right to that good (Malek, 2002).

CONCLUSION

Government passes legislation preventing price gouging based on the implicit premise that it can allocate resources more efficiently than the market. By doing so, it alleges that it knows what the people want better than entrepreneurs who sink or swim based on their ability to anticipate matters of this sort.[9] Government regulation is nothing short of a disaster as far as satisfying customers is concerned.

Paradoxically, the government is correct in thinking that the people want "fair" prices for goods. Were this not the case, they would not dare to oppress price "gougers." Indeed, this is a wildly popular stance for a politician to take. But, the masses of the people are entirely erroneous in supposing that these market interferences can actually be helpful to them. In the words of Caplan (2007), "the public suffers from anti-market bias."

Particularly, during times of disaster, prices must be allowed to adjust as a signal to producers and consumers alike. Consumers will utilize less of these goods, and producers will increase their output. As the supply adjusts following the price increase, goods and services will get to those who want them the most and are willing to pay for them. This will undoubtedly be a more effective way to distribute supplies to hurricane victims. The government has demonstrated on more than one occasion how ineffective it can be, and this unwarranted attack on price gouging is no exception. Price controls must be repealed, and the free market must be allowed to work via the beneficent invisible hand[10] (Smith, 1979; Childs, 1977; Diamond, 1987; Flew, 1987; Garrison, 1985), not by the stultifying hands of the bureaucrats and politicians.

NOTES

1. They had access to the Sears Roebuck catalog, and to financial newspapers of the west. There was also that slight matter of the black market, which, according to many estimates, was significant.

2. The literature supporting this statement emanates from the Austrian side of the socialist calculation debate (Boettke, 1991; Gordon, 1990; Hoppe, 1996; Klein, 1996; von Mises 1981; Rothbard, 1991).

3. That is, unless we define, Marxist-like, any seller of anything, even labor, as a capitalist exploiter. Remember, after Katrina, wages in New Orleans catapulted.

4. Sometimes, what prevents markets from operating in this way are riots on the part of the populace, not formal governmental decrees. See on this: http://www.labournet.net/world/0010/zimbab1.html www.labournet.net/world/0010/zimbab1.html; http://select.nytimes.com/gst/abstract.html?res http://select.nytimes.com/gst/abstract.html?res = FA0717FE3D580C738EDDAD0894D1484D81&n = Top%2fReference%2fTimes%20Topics%2fOrganizations%2fI%2fInternational%20Monetary%20Fund%20. It cannot be denied of course, that in these cases the masses of people are acting in a governmental manner: seizing property that does not belong to them.

5. As Hugo Chavez, the Venezuelan dictator is now discovering, price controls, of which "anti-gouging" is one aspect, is a recipe of clearing goods off of store

shelves. How this will help consumers is a bit difficult to ascertain. See on this: http://blog.mises.org/archives/006271.asp.

6. On May 23, 2007, the House of Representatives passed an anti-oil price gouging bill by a vote of 284–141. www.latimes.com/news.nationworld/nation/la-na-gasoline24may24,03926741.story?coll=la-home-nation: www.washingtontimes.com/business/20070524-121323-8807r.htm.

7. If he does this, however, it will only underscore even the more, how far we have come from constitutional principles. Can anyone imagine George Washington or John Quincy Adams acting like a nanny, and urging citizens to consume fewer apples? For a principled constitutional outlook, see: Ron Paul (www.mises.org/story/2514).

8. For an Austrian economic analysis according to which such institutions are necessarily part of the system of government privilege not free enterprise (Anderson et al., 2001; Armentano, 1999; Block, 1994; Boudreaux and DiLorenzo, 1992; McChesney, 1991; Rothbard, 1970; Shugart, 1987; Smith, 1983).

9. When was the last time a government bureaucracy went broke due to its inability to serve the public? Never, that is when. It is for this reason, perhaps, that one of the most popular bumper stickers in the New Orleans area articulates this message: "FEMA happens."

10. "But the annual revenue of every society is always precisely equal to the exchangeable value of the whole annual produce of its industry, or rather is precisely the same thing with that exchangeable value. As every individual, therefore, endeavors as much as he can both to employ his capital in the support of domestic industry, and so to direct that industry that its produce may be of the greatest value. Every individual necessarily labors to render the annual revenue of the society as great as he can. He generally, indeed, neither intends to promote the public interest nor knows how much he is promoting it. By preferring the support of domestic to that of foreign industry, he intends only his own security; and by directing that industry in such a manor as its produce may be of the greatest value, he intends only his own gain, and he is in this, as in many other cases, led by an *invisible hand* to promote an end which was no part of his intention. Nor is it always the worst for the society that it was no part of it. By pursuing his own interest he frequently promotes that of the society more effectively than when he intends to promote it. I have never known much good done by those who affected to trade for the public good. It is an affection, indeed, not very common among merchants, and very few words need be employed in dissuading them from it" (emphasis added by present authors).

Section VII

POPULATION

Chapter 17

Armageddon?

This chapter argues that the belief that the rapid growth of the human population will inevitably lead to a major ecological disaster is neither intuitively nor empirically tenable. A significant portion of the world's pollution comes not from overpopulated poorer nations, but from Western nations with very low population growth rates. On the other hand, most of this damage is the result of misguided government policies, and not Western overconsumption. Overpopulation is not likely to be a problem environmentally because Malthusian predictions are often made based on the assumption that current rates of resource use and population growth rates will remain the same. Such assumptions ignore the critical role that adaptability has played in allowing humans to avert Malthusian crises. Substitution of products, innovative production methods, and technological changes all ensure in the long run, very few of the facts that predictions are based on will remain fixed.

INTRODUCTION

One of the most pressing issues in environmentalism today is overpopulation. Population control advocates (PCAs) believe that the unprecedented growth in the number of humans on Earth has been the cause of numerous ecological problems, including the destruction of the rainforests and the hole in the ozone layer. PCAs argue that unless drastic action is taken to curb the population growth rates, especially in the Third World, major environmental catastrophes will take place. They argue in favor of massive birth-control programs targeted toward poor nations, which in practice often violate human rights.

There are many problems with these arguments. No causal relationship between rapid population growth and environmental degradation is shown, as most pollution comes from industrialized nations. The converse argument that Western overconsumption is responsible for environmental degradation is also not true, as human suffering and pollution are often caused by misguided government policies and not free markets. In reality, there is no crisis, since the Earth has the ability to support its growing number of human inhabitants. The predictions of resource exhaustion are based on flawed arguments of diminishing returns, which are overcome by technological innovation, and the pressures of the price system.

Roughly about six billion people now live on Earth. For most of history, the population remained stable; from 1650 to 1930, it grew from 500 million to about 2 billion. Even this rise was substantial historically, as it was the result of the Newtonian Revolution in science which allowed humanity to manipulate the environment to lower infant mortality rates and to increase life expectancy. Since that time, technology and science have contributed to the extremely rapid growth of our numbers. Sixty years ago there were two billion people; since then the total has risen by 200 percent, and is increasing currently at a rate of over 95 million per year (Fornos, 5). Although world birth rates have dropped to 1.8 percent from a peak of 2.1 percent in the mid-1960s, mortality rates have also fallen, and the number of humans is expected to reach over 10 billion by the middle of the twenty-first century (Ehrlichs, 16).

THE POPULATION CONTROL HYPOTHESIS

PCAs have interpreted the preceding data as the cause of most of the world's current ecological problems. According to these theorists, humanity has exceeded the "carrying capacity" of the Earth, and is now beginning to pay the price, one that will steepen in the coming years. According to Paul and Anne Ehrlich, *overpopulation* is defined as excessive "numbers of people in one area relative to its resources and [exceeding] the capacity of the environment to sustain human activities" (38). The destruction of 750,000 acres of Brazilian rainforests in a week and the massive famines that result in the deaths of 42,000 children a day are taken as the consequences of exceeding the Earth's carrying capacity (Fornos, May/June 5; November/December 31). These "doomsters" argue that population growth is directly responsible for environmental damage, in particular the destruction of the rain forests, the hole in the ozone layer, and the greenhouse effect. Since humans are increasingly reliant on natural resources and prolific crop yields for their survival, excessive numbers of people will put severe strains on the Earth, depleting energy reserves and converting the precious homelands of numerous plant species into land for habitation.

The problems are considered to be extreme and therefore require drastic solutions. For example, the use of Norplant as a female hormonal contraceptive has been rampant in Third World birth-control clinics.[1] The women who are injected with the contraceptive are rarely told of the dangerous side effects, which include strokes and kidney problems (Smith, 43). In fact, over 50 percent of women who have the device implanted want it removed after two years (Smith, 46). Even more questionable is the policy of the International Centre for Diarrhoeal Disease Research in Bangladesh which denies healthcare to needy women in order to "induce women to, accept contraceptives" (Smith, 58). In some cases women have been sterilized without their permission. The coercion is considered justified because it is ostensibly in the best interests of the world environment, even to those who are coerced into having sterilization procedures performed. Garret Hardin writes with admiration of the pressure put on Chinese women to have abortions:

> Such coercion in the East should be compared to forcing a Westerner to pick up the litter he or she has dropped on the ground at a public park . . . the environment is seen as the possession of the group; littering it (with anything) is not a right of the individual. (12)

If children are considered as "litter," then there is no limit on what solutions can be justified to solve the problem of overpopulation. Clearly, violating human rights and denying people the right to make free choices about whether or not to have children are within the range of possibilities for population planners.

AN EMPIRICAL REFUTATION

The truth is, however, that the Earth is not overpopulated. For one thing, no direct causal relationship between rapid population growth and environmental degradation has ever been clearly shown. For doomsters, the relationship is merely assumed to exist. Since humanity's rapid growth is occurring at the same time as massive damage to the environment, one must therefore be the cause of the other. Statements such as the following are common in the disaster literature:

> Since [1975] our reliance upon these gases [CFCs] has increased by 100 percent. We do not really know how much damage we have caused to the atmosphere.[1]

Meanwhile, the world's population of 5.3 billion is expected to reach 6 billion by the middle of the decade [1990s], and, at the present rate, it will double within forty years. (Fornos, May/June 5)

The statement simply assumes that in the "meanwhile," the increase in the number of humans on the planet has led to a rise in the production of CFCs, but never explicitly shows this. One of the best refutations of Malthusianism has been provided by Sowell (1983). He reasons that if overpopulation will lead to immiseration in the future, the cross-sectional data ought to be at least compatible with this time series prediction. That is, if greater numbers will lead to poverty, starvation, and death in the future, we ought to be able to discern such a pattern now. In order to see if this is so, this economist arranges countries on a grid, showing the relationship between a nation's wealth and population density (see page 203).

If the Malthuses and the Ehrlichs of the world were correct, we ought to see highly dense and poor countries at one end of the spectrum (i.e., El Salvador, India, and Bangladesh), and low-density rich ones at the other end (Sweden, Nigeria, the United States). And indeed, these two boxes are not empty. But, no discernible upward and to the right pattern emerges from this analysis, as implied. Instead, there are numerous countries which go against the over-populationist grain. There are those which are densely populated and rich (West Germany, Belgium, Holland, Japan) and those which are relatively empty and very poor (Columbia, Algeria, Chile). Thus, those who cite large numbers of human beings as the cause of human difficulties will find no comfort in the facts of the matter.

The fact is that a relationship between environmental degradation and burgeoning Third World populations does not exist (Block, 19). The truth is that over 75 percent of the world's pollution comes from the industrialized rich nations of the West (Smith, 50). Few of the truly poor in the South[2] can even begin to approach the damage done by misguided government policies of many Western, and, Third World nations. For example, "the cutting of trees in the Amazon is not related to an excess of peasants cutting trees for survival but rather the government policy of supporting commercial ranching" (Smith, 50). The world's extreme poor, who contribute to 90 percent of the growth in number of the Earth's inhabitants, can hardly be blamed for the production of CFCs or the dumping of toxic chemicals into the oceans or on public lands.

GOVERNMENT, NOT MARKET, DEPREDATION

The dangerous conclusion is that Western "overconsumption" is the cause of pollution. This side in the population debate argues that Western industry exploits Third World nations for their raw materials, and has forced the migration of poorer people into rain forests and other fragile ecological environments. The solution would be to punish the industry and place limits on its use of the resources of underdeveloped nations.

This argument is not tenable either. The destruction of the Brazilian rain forest is not due to private industry ownership of the land, but to the government's policy of giving the hind away for commercial use. Another example of the recklessness of governments is given by Lester Brown. He discusses the policy in Britain under Margaret Thatcher of privatizing the energy sector of the economy. However, "no investor in a market situation would invest a dollar in nuclear power plants . . . because they are uneconomic . . . [industry] would not touch them" (28). Nuclear power, as evidenced by the Chernobyl disaster, is an extremely dangerous form of energy. Governments are free to experiment with it because they, unlike business, do not have to take into account the costs and benefits, as well as the economic liability for potential harm. Brown discusses the ecological damage that has been done in Poland, the former USSR, Czechoslovakia, and Mexico. Yet, these nations are hardly shining examples of free market capitalism.[33] Western industry, when free of government intervention, does a much better job of controlling pollution and ensuring that resources are not depleted because it must take into account market pressures. If a firm were to recklessly pollute, it would be held liable in court for the damage done to other's property; if it wasted precious natural resources, it would be driven out of business by competitors who were more careful in their use of the resource.

TRAGEDY OF COMMON OWNERSHIP

The ecological damage the world experiences is not the result of Western industry's overconsumption of raw materials. Even though it pales in comparison to the damage done by governments to the environment, the use of wood by peasants as a major source of fuel is a problem. The practice of slashing and burning forests to support cash crops has led to soil erosion and deforestation. Still, the erosion of topsoil in the Himalayas and the destruction of African and South American rain forests is a classic example of the tragedy of the commons. Peasants are pushed off their land by wars and government policies, and turn to the use of free, unowned land in order to survive. Vice President Gore notes in his book *Earth in the Balance* that the genetic treasure map of the world, where indigenous species live, is being erased. The land is being cleared away for use by peasants and governments. He argues, "These fragile places serve as the nurseries for and reservoirs of genetic robustness, vitality and resilience, but all of them now are in serious jeopardy" (128).

Without them, plants will succumb to disease and blights, and humanity will have no way to combat plant diseases resistant to pesticides.

The truth is, however, that the lands in which these plants[44] grow are unowned, and they are being destroyed for the same reasons as in the case of the rain forests. Actually, the massive search for disease-resistant strains of plants is often funded by agribusinesses, demonstrating the responsibility and commitment of private industry to resource preservation. The problem is not the result of too many humans destroying fragile lands in the search for a place to live, but the lack of private ownership of the land and the reckless policies of governments that has exacerbated environmental problems.

RESOURCE DEPLETION?

The major flaw in the doomsters' argument is a misunderstanding of the Earth's ability to sustain burgeoning populations. PCAs mistakenly make predictions about the future based on present rates of consumption, resource depletion, and population growth. They do not consider changes in trends or the impact of scarcity in developing alternative methods of production and the use of substitutes. The first flawed argument is that the Earth's "carrying capacity" has been exceeded—humanity has outgrown the land's ability to support adequate living standards. This theory states that population density per se is not a problem so much as the enormous quantity of resources consumed by Western nations in comparison to the actual resources of the land they live on and their proportion of the Earth's total population. In fact, "20 percent of the world's population [uses] 80 percent of the world's resources" (Fornos, November/December 30).

The argument is that if the current consumption patterns of the West are emulated by the developing nations, the world will quickly run out of resources. The Ehrlichs argue that resource-poor nations such as Japan or the Netherlands have been able to support population densities of close to 1,000 people per square mile "only because the rest of the world cannot . . . the people of the Netherlands didn't build their prosperity on the bounty of the Netherlands, and are not living on it now" (39–40). The assumption is that there is a finite quantity of resources that is being proportionally overused by tiny industrialized nations. In order to raise the living standards of underdeveloped nations, the consumption of the West must be reduced, or else the world's resource pie will be cut into smaller and smaller pieces.

This argument ignores the benefits of specialization. The Netherlands gained its riches precisely because it utilized its scarce resources and traded with other nations for the resources it needed. As Julian Simon argues, trade is a mutual exchange of goods, not exploitation: "Saudi Arabia no more 'supports' the Netherlands by exporting oil than the Netherlands 'supports' Saudi Arabia by exporting electronic goods" (271). Specialization has been one of humanity's greatest achievements because it allows people to overcome dependence on their own talents and resource endowments. In New York City, for example, few people have any knowledge of how to milk a cow, yet millions of gallons of milk are consumed there every day. In the Great Plains, farmers have lived for hundreds of years without trees to build houses or to warm homes in the winter. To argue that countries are overpopulated because they cannot support themselves would be to ignore the fact that this has been the case for hundreds of years, and that it actually has resulted in improved standards of living. Indeed, life per capita income, and life expectancy is often by rule greater in cities than in rural areas (Simon 249–253).[5] The fact is that there are typically more benefits than costs from "overcrowding."

MALTHUS REJECTED

Ultimately, the major problem with the argument is that it is based on Malthusianism. In the early 1800s he predicted a dismal future awaiting mankind: crop yields would not be able to keep up with the rapidly growing populations and that massive deaths due to starvation would occur. Malthus was wrong, however, because while "population would grow geometrically . . . he didn't foresee our ability to make geometric improvements in agricultural technology" (Gore 127). Malthusian thought nevertheless lives on in the writings of the Ehrlichs, Werner Fornos, and many others. Their belief is summed up by statements such as this from the Ehrlichs: "At its current growth rate, the population [of the Philippines] will double in twenty-five years. Agricultural production in the Philippines is not likely to double in the foreseeable future; indeed, it may even decline" (75).

The essential factor ignored is the crucial role played by scarcity and innovation. They can radically alter the outcomes of predictive models. The problem is that all things do not remain equal. For example, the Ehrlichs argue that the catches of fishermen are declining to dangerously low levels relative to the world population, and that soon the overfishing (the "vacuuming of the sea" as they call it) taking place in the oceans will render them barren (Ehrlichs 85). They do not seem to realize the implications of the historical data they present in making the implications of the historical data they present in making this argument. In 1970, 13 million tons of Peruvian anchovies were caught, but "the Peruvian anchovate fishery collapsed . . . [and] the fisheries harvest [dropped] to just over two million tons in 1973." By 1986, however, the catch had risen to over 7 million tons, a "sharp upturn." What happened to so dramatically change the fortunes of the Peruvians? They more aggressively pursued the fish, and new search methods in new areas allowed the total world catch to rise in the same period from 70 million to 84 million tons (Ehrlichs 85–87).

This "vacuuming of the sea" has yet to be manifested because it is based on a misunderstanding of the law of diminishing returns. The theory posits that the productivity of anything should decline if the total stock of the resource used (in this case the fish) rises and the production factors, capital and labor (boats and fishermen), are held constant. Therefore, costs will rise as more and more inputs must be used in order to maintain a given level of productivity. Ultimately, the costs will rise to the point that it is no longer economically feasible to use the resource, or the resource will be exhausted by overuse. The law of diminishing returns is only true, however, in the short run. In the long run, production factors do not necessarily remain constant, and it is not true that the stock of a resource (especially a renewable one such as fish) is fixed, and that use only draws down the stock. New production methods, and innovative technology help increase productivity. Actually, growing populations contribute to gains in productivity that make labor more abundant, lowering its cost and better utilizing its production capabilities. The fact is that as while the world has been experiencing its so-called Malthusian crisis, the production of food globally has out-paced population growth by over 1 percent since the 1940s (*National Review* October 18, 1996, p. 18). The problem for the Neo-Malthusian tradition is that technological gains and innovations can never be predicted, but worst-case scenarios based on present facts often seem appalling. Thus, it always appears that the world is running out of resources and is reaching unsustainable levels of production.

THE FREE MARKET PRICE SYSTEM

There is a second Malthusian flaw: it fails to incorporate the impact of scarcity and prices on incentives. Julian Simon mentions the erroneous predictions of nineteenth-century English social scientist W. Stanley Jevons that Great Britain was going to run out of coal, and its industry collapse. Instead, the United States has developed enough coal reserves to supply England for the next 1,000 years, and that country England has numerous oil and natural gas reserves (Simon 93). This has occurred because the price of coal increased as it became scarce, and suppliers, driven by the lure of potential profits, put more effort into finding reserves and developing more efficient ways to harness its energy. In addition, demand for coal decreased as alternatives such as oil and natural gas became more attractive. The potential profits in these sectors rose as well, and a rush of new suppliers helped to lower the costs of these alternatives, making them more viable. Ultimately, coal never did run out. Indeed, an argument can be made that we used coal too slowly, not too quickly.

Modern-day Malthusians fail to take the price system[6] into account, instead relying on "technological rather than price-trend data in their forecasts" (Simon 101). If a good becomes scarce in a free market, its price will rise. Substitutes will be used, as well as new production methods, and an equilibrium will be established. This tends to preclude the exhaustion of the resource. This also applies to the destruction of crops by viruses. If this occurs, the price of the plant will rise, and consumption will decline. People will not starve, however. More attention will be focused on finding disease-resistant strains, and consumers will use substitutes. In general, this is the problem with all theories of overpopulation—they fail to take into account that behavior alters conditions. Humanity is not doomed simply because its numbers are growing rapidly. Clearly, when resources become scarce and productivity declines, changes in consumption and production are made, and life goes on, albeit somewhat differently, though by no means worse than before.

NOTES

1. Simon describes the brutal mistreatment of women by birth-control clinics, and the gross disregard of human rights. See page 43.

2. The South is a term commonly used to refer to the lesser developed nations in the world, as opposed to the more industrialized nations of Europe, Japan, and North America. The dichotomy is not geographically accurate since poor and rich nations are by no means aligned in the exact same way, and it has come under some criticism for being a provincial term used by more developed nations. Still, it is reasonably accurate, and therefore it is used to distinguish poor and rich nations in this chapter.

3. While Brown's criticism is not limited only to Communist nations, command economies such as the one he mentioned are much more likely to experience severe environmental damage than Western nations. See Gore, 248: "It is no accident that the worst environmental tragedies were created by communist governments."

4. Every crop in the world is derived from certain plant species, which originate from specific geographical genetic centers. There are only twelve such centers in the world, each with the genetic home of about twelve species each. While the species can be transported and grown in other regions, they lack genetic diversity that comes from the wild indigenous growth of the species in its native environment. Since most of the large crops in the world contain plants that are genetically homogenous, they often are susceptible to viruses and blights. It is necessary then to find a disease-resistant variation of the plant to protect the crop. The problem is that the destruction

of so much of the genetic homelands has made finding disease-resistant variations extremely difficult (Cf. Gore 132).

5. Simon uses statistics from Great Britain over the last two hundred years to show the impact of technology on the quality of urban life. In 1841, the life expectancy in the United Kingdom was 40, while it was only 35 in the city of London. Since the technology of waste disposal and sanitation practices improved, cities now have high correlations with higher life expectancies. For example, in London, just after World War II, the average life expectancy in London surpassed the rest of Britain for the first time; 67.3 compared to 66.4 (Simon 245–249).

6. After World War II, the president's Materials Policy Council argued that raw materials were not going to be able to keep pace with the higher demand of the previous few years, and the prices of resources were going to rise substantially. Instead, most of the materials it studied declined in price in the next twenty-five years. This illustrates the mistakes that can be made when forecasts are made on the basis of short-term variations in use, as well as ignoring long-term historical evidence (Simon 35).

Chapter 18

Cause of Poverty?

POVERTY MISUNDERSTOOD

People are confounded by the presence of poverty in a world in which some individuals and nations amass unbelievable amounts of wealth. Unfortunately, many people fail to understand the "poverty problem" because they incorrectly believe that wealth is the norm and poverty the exception.

The reality, however, is that it is the presence of wealth—rather than its absence—that requires explanation.[1] Wealth does not occur spontaneously in nature, yet much of the debate on this issue centers not on the origins of wealth but on the poorly formulated question of how poverty came to be.

Population growth, for example, is often cited as a cause of poverty in the world. Scientists such as Dr. Paul R. Ehrlich, groups such as the Hugh Moore Fund, and even the presidential report *Global 2000* routinely sound the population alarm as the bane of poor countries and soon the whole planet. Such spokespeople attribute everything from droughts, damaged crops, famines, deforestation, pollution, international conflict, epidemics, gridlock, drug use, and crime to overpopulation.[2] They claim our increased numbers are not only the chief cause of poverty but also the crux of many maladies afflicting civilization today. Problems such as racism, sexism, religious discrimination, warfare, and environmental deterioration cannot be solved until the population question is addressed. Says Ehrlich, "Whatever your cause, it is a lost cause without population control."[3]

The scare is by no means new. It started 200 years ago with perhaps the oldest and best-known population alarmist Sir Thomas Malthus, who in 1798 wrote the famous *Essay on the Principle of Population.* He believed that, given the human tendency to eat and reproduce, the world would experience exponential population growth but only arithmetical food growth. As soon as population levels exceed food levels by some critical amount, a Malthusian dilemma occurs. These occurrences are said to be responsible for the near annihilation of the inhabitants of Easter Island and the loss of over 20 percent of Europe's population during the Black Death. Such dilemmas are inherent in human activities, Malthusian theory states, unless societies actively work to lower their birth rates.

The trouble with Malthusian doctrine is that it is prescriptive rather than descriptive of history. When Malthus wrote at the end of the eighteenth century, empirical evidence on the subject of population was nearly non-existent. A national census had yet to be taken in Great Britain, and for the less developed parts of the world, precise numbers are still hard to determine today, let alone 200 years ago.[4]

As the nineteenth century progressed and more census data became available, we learned that the food supply was actually growing faster than the population. This trend has continued into the twentieth century.[5] Even with population many times greater than it was 100 years ago, food per capita is generally higher today.[6] Malthus was certainly on track when he recognized the geometric growth pattern of population, but he failed to see our ability to make even greater geometric improvements in agricultural technology. History has shown that as population increases, so does our ability to grow and distribute food.[7]

MISSING EVIDENCE

In the 200 years since Malthus, the argument has changed little; doom is still on the horizon. Grave famines and tremendous scarcity are predicted for every decade. The price of food is scheduled to increase and resources to decrease. Amazingly enough, these extrapolations never materialize. The average person is healthier, wealthier, and eats more nutritiously than he or she did 30 years ago. Life expectancy has increased as infant mortality has decreased, and food production is still rising at a faster rate than population.[8] The only thing that proponents of overpopulation have correctly predicted is that population grows.

Despite the overwhelming empirical evidence to the contrary many self-styled environmentalists, including Vice President Al Gore, hold steadfast to the notion that overpopulation is the cause of nearly all of our ailments, especially poverty. Such critics advocate that the United States and the United Nations adopt a plan of zero population growth. Time and time again, these people are proven wrong, both logically and empirically; low population density is not linked to the well-being of any nation.

The economist Thomas Sowell, in his book *The Economics and Politics of Race*, devised a powerful way to understand the lack of correlation between population density and poverty. Sowell reported on 61 countries with regard to these two variables (Table 18.1).[9] If the overpopulation argument is correct, then the countries with the highest incomes should have the least population and the poorest countries should have the most. If the overpopulation argument is wrong, then there should be no tendency for any nations of any particular wealth to concentrate in any one population category. Table 18.1 indicates that a wide distribution of nations exists throughout the different population and income classifications.

Sowell points out that Ethiopia has nearly the same population density as the United States, but averages less than $100 in annual per capita income.[10] Conversely, Singapore has more than 10,000 people per square mile and a per capita income that is 200 times greater than that of Ethiopia. Japan, one of the richest of the non-European countries, has a higher density per square mile than does India, one of the poorest nations in the world.[11]

Table 18.1 Population Density and Income

PER CAPITA INCOME		*POPULATION PER SQUARE MILE*		
(Nations in each column listed in order of income)		500 or more	100–499	Less than 100
	More than $5,000	West Germany	Kuwait	Sweden
		Belgium	Denmark	United Stales
		Holland	France	Nigeria
		Japan		Canada
				Australia
				Austria
				Libya
				Saudi Arabia
	$1,000 –$5,000	Great Britain	East Germany	USSR
		Singapore	Spain	Iran
		Barbados	Israel	Iraq
		Jamaica	Greece	Argentina
		Taiwan	Italy	South Africa
		Lebanon	Ireland	Brazil
			Poland	Mexico
			Costa Rica	
	Less than $1,000	El Salvador	Turkey	Colombia
		India	Cuba	Algeria
		Bangladesh	Malaysia	Chile
			Morocco	Guyana
			Philippines	Bolivia
			Egypt	Liberia
			Indonesia	Congo
			China	Tanzania
			Uganda	Kenya
			Pakistan	Afghanistan
			Burma	Ethiopia

Source: *The World almanac and book of Facts,* 1981.

Close analysis of the table lends no evidence to the overpopulation argument. The countries contained in the upper left cell have more than five times the per capita income of those in the lower right cell and also have more than five times as many people per square mile.[12] The opposite is also true. The nations in the upper right cell have more than five times the per capita income of all the nations in the lower left cell, coupled with less than one-fifth as many people per square mile.[13] There is no correlation between population density and wealth.

UNDERSTANDING POVERTY

What, then, accounts for poverty? The Fraser Institute, in conjunction with research institutes from 10 other countries, including the Cato Institute of the United States and the Institute of Economic Affairs of England, released a study that strongly suggests economic freedom determines a country's well-being. James Gwartney, Robert Lawson, and Walter Block give a "freedom ranking" to over 100 countries in the new study entitled *Economic Freedom of the World: 1975–1995*.

According to the authors, the central elements of economic freedom are personal choice, protection of private property, and freedom of exchange.[14] They state that individuals have economic freedom when they acquire property legitimately and when that property is protected from illegitimate invasion or seizure. Free individuals may do with their property whatever they desire as long as they do not violate the identical rights of others. In such a society the fundamental role of government is to protect private property and to enforce contracts made between consenting individuals or firms. When government exceeds this limited capacity, economic freedom declines.[15]

The study uses 17 measures grouped into our categories to rate economic freedom. Over 100 countries are assigned a rating from 0 (no freedom) to 10 (complete freedom).

The four basic categories are:

Money and inflation. This includes the government's ability to protect money as a store of value so that it can be used as a medium of exchange. The measure also includes the volatility of inflation, monetary growth relative to the potential growth capacity of an economy, and the right of a citizen to own foreign-currency accounts at home and bank accounts abroad.[16]

Government operations and regulations. What role, if any, does the government play in the decisions of production and consumption? Criteria include public spending as a share of gross domestic product (GDP), the size of the state-controlled sector, absence of price controls, freedom to enter markets, and controls on borrowing and lending rates.[17]

"Takings" and discriminatory taxation. Are citizens and firms free to earn and to keep their wealth? Considered are subsidies and transfer payments as a share of gross domestic product, the level and impact of marginal tax rates, and absence of conscription.[18]

International exchange. How free are individuals to exchange goods and money with foreigners? Variables include taxes on international trade, differences between an official exchange rate and a black market one (a measure of exchange controls), the size of a nation's trade relative to the size that might be expected (an estimation of non-tariff barriers), and restrictions on capital flows.[19]

While there is no relation between population density and wealth, certainly a strong one exists between economic freedom and wealth. This was especially evident for countries that sustained a high level of freedom for many years. The study suggests that individuals and firms will not respond positively to economic freedom unless they believe it is likely to last. It is no surprise that the six countries that had persistently high ratings throughout the 1975–1995 period (Canada, Germany, Hong Kong, Singapore, Switzerland, and the United States) are also in the top 10 for per capita GDP in 1993–1995.[20]

No nation with a consistently low rating was able to achieve even a middling income. In the 16 countries for which the index fell the most during the 1975–1995 period, average real GDP per capita fell at an annual rate of 0.6 percent.[21] They are Algeria, Brazil, Congo, EI Salvador, Greece, Honduras, Iran, Morocco, Nicaragua, Panama, Sierra Leone, Somalia, Syria, Tanzania, Venezuela, and Zambia.

Of particular note are the ratings given to the fast-growing economies of East Asia and Japan. It is commonly assumed that economic freedom is restricted in these countries.[22] According to Gwartney et al.'s report, however, Singapore ranks second out of the 103 countries in economic freedom. Malaysia is sixth, Japan ninth, and South Korea twelfth. Many people confuse economic freedom, which is present in these parts of the world, with political freedom, which is lacking.

Most notable is what happens to a nation's GDP as it moves up the "freedom scale." A one point increase in economic freedom will raise the average country's growth rate by one percentage point. This does not sound astounding until the effects of compounding over time are considered. For example, a growth rate of 2.5 percent a year raises GDP by 28 percent at the end of 10 years. In contrast, a growth rate of 3.5 percent year is associated with a GDP increase of 41 percent in a decade.[23]

In the *Economic Freedom of the World*, each country is assigned a letter grade corresponding to its level of economic freedom. Countries that scored an A had a per capita GDP average of $15,834, while those that received an F– had a per capita GDP average of $1,650. See Table 18.2.

To bring the findings from the *Economic Freedom of the World* full circle, it is necessary to recreate and enhance the table Thomas Sowell used to discredit population pressure theory. Table 18.2 is similar to table 18.1, except that the vertical distance measures per capita income instead of real per capita GDP. The nations used in this table vary somewhat from those used by Sowell.[24]

The number that appears in the top right corner of cells is the average "economic freedom rating" for all the countries contained in that cell. The letter after the number indicates the average letter grade for the countries in that cell. Ratings of 9.0 or more are equivalent to an A+, 8.0 or more to a A, 7.0 or more to a B, 6.0 or more to a C, 5.0 or more to a D, 4.0 or more to a F, and 3.99 or less to an F–.[25]

Just as in Sowell's table published 15 years earlier, no relationship exists between population density and wealth. Of the three cells with the highest population density, the nation that has the highest rating overall (6.8) is also the richest. Further, those nations with the highest per capita GDP also have the highest economic freedom rankings and are evenly distributed across all three population densities. Inversely, the countries with the lowest per capita GDPs also have the lowest economic freedom rankings and are distributed across the three different population densities.

Satisfying the Critics

Some critics find this "economic freedom argument unsatisfying"; they claim it does not deal with quality of life issues for all members of society, especially those who do not partake in economic expansion. To address this issue, the United Nations published the *Human Development Report* in 1993 to rate countries on a scale that included more than just economic factors. The report was authored by Mahbub ul-Haq, an economist and former Pakistani finance minister, who designed a Human Development Index to rank countries from 0.000 to 1.000 in what amounts to quality of life.

The United Nation's goal was to recognize countries that translate national wealth into higher living standards for their citizens. It assumed that economic growth should allow individuals to lead fuller lives.[26] Ul-haq stresses

Table 18.2

	EXHIBIT					
Grade	*A*	*B*	*C*	*D*	*F*	*F–*
Per Capita GDP ($) 1994	15,834	13,569	7,888	3,784	3,068	1,650

Source: *Economic Freedom of the World: 1975–1995*, xxii

Table 18.3 Population Density and GDP

REAL PER CAPITA GDP (Nations in each column listed in order of income)		*POPULATION PER SQUARE MILE*		
		500 or more	100–499	Less than 100
	$10,000 or more	**6.8/C** Singapore United Kingdom Japan Taiwan Netherlands Germany Belgium	**6.1/C** Switzerland France Denmark Austria Italy	**6.6/C** New Zealand United Stales Canada Australia Norway Finland Sweden
	$5,000 – $9,999	**5.7/D** South Korea Mauritius Israel	**5.8/D** Malaysia Mexico Spain Portugal Greece	**5.5/D** Argentina Chile Venezuela
	Less than $5,000	**5.35/D** Philippines India	**4.85/F** Thailand Costa Rica Guatemala Indonesia Honduras Portugal Czech Rep. Poland Kenya Egypt Syria Ghana Turkey Nigeria Iran	**4.2/F** Bolivia Peru Botswana South Africa Cameroon Tanzania Zimbabwe Nicaragua Brazil Zaire

Source: *Economic Freedom of the World: 1975–1995* and *The World Almanac and Book of Facts 1995*

the fact that many countries experiencing growth do not necessarily see a rise in employment that would allow more people to participate in the prosperity.

The UN report concentrates on very different criteria than the *Economic Freedom* study. Prime components are life expectancy, educational standards, and individual purchasing power. More particular measurements include adult literacy rates, mean years of schooling, and access to basic needs such as health services, safe water, and sanitation services.

Table 18.3 indicates, just as the previous tables do, that population density has very little bearing on anything significant. The table is constructed in the same manner as Table 18.4, using the same countries. The number on the top

Table 18.4 Population Density and GDP

Real Per Capita GDP (Nations in each column listed in order of income)	*POPULATION PER SQUARE MILE*		
	500 or more	100–499	Less than 100
$10,000 or more	**0.943** Japan Netherlands United Kingdom Belgium Israel Singapore	**0.951** Switzerland France Germany Denmark Austria Italy Spain	**0.969** Canada Norway Sweden United States Australia Finland New Zealand
$5,000 – $9,999	**0.795** Mauritius	**0.848** Greece Czech Rep. Portugal Mexico Malaysia	**0.844** Chile Venezuela
Less than $5,000	**0.475** South Korea India	**0.573** Costa Rica Poland Turkey Indonesia Guatemala Egypt Kenya Pakistan Ghana Nigeria	**0.504** Argentina Brazil South Africa Peru Iran Botswana Nicaragua Honduras Zimbabwe Bolivia Tanzania Cameroon Zaire

Sources: *Human Development Report 1993* and *The World Almanac and Book of Facts 1991*

right corner of each cell is the average Human Development Index rating for all the countries contained within the cell. The table indicates that the Human Development Index (or quality of life) does not decrease as population density increases.

The three cells that received the highest quality of life rating are in the top rows, and the countries are evenly distributed among the various population densities. The three cells with the lowest rating are all shown in the bottom row, and once again are unaffected by population size.

A comparison of tables 18.2 and 18.3 shows remarkable similarities. Six of the top 10 countries are shared by both rating systems: Australia, Canada, Japan, Switzerland, the United Kingdom, and the United States. Four of the bottom 10 countries are shared by both tables: Egypt, Nigeria, Tanzania, and

Zaire. Comparison of the two surveys supports the notion that economic freedom, economic development, and quality of life are closely related.

The *Human Development Report 1993* does more than rank countries. It also suggests methods for countries to improve their own development and ways for the international community to aid them. The report also deals with the major obstacles preventing countries from improving their quality of life. The four prime factors are poor and arbitrary legal systems, bureaucratic restraints, maldistribution of assets, and outdated social norms.[27] The first three are very similar to those mentioned in the *Economic Freedom of the World 1975–1995*. Nowhere in the 230 pages of the report is population density listed as an obstacle to developing a higher quality of life.

Mysterious Popularity

With no evidence to support the notion that population is linked to the wealth or living standard of a nation, it is a mystery how the argument for overpopulation can retain its popularity. One explanation is that imposing population controls upon a populace in order to reach the "correct" density offers a simple short-term solution to the complicated problems that countries face today.

Consider the People's Republic of China, a nation widely claimed to be overpopulated. With only 288 people per square mile, it has led the way in controlling population.[28] If China is able to severely restrict population growth for the next 30 years, none of its problems will actually be solved. China's problem is its controlled economy. With fewer people but equally onerous controls, it will never overcome its population and consumption dilemmas. China will one day discover that the only correlation between population density and poverty is the fallacious nature of the overpopulation argument.[29]

NOTES

1. Thomas Sowell, *The Economics and Politics of Race* (New York, NY: William Morrow and Company, Inc., 1983), 208.
2. Paul R. Ehrlich, and Anne H. Ehrlich, "The Population Explosion: Why Isn't Every one as Scared as We Are?" *The Amicus Journal,* Winter 1990, from Matthew Polesetsky, ed., *Global Resources* (San Diego, CA: Greenhaven Press, Inc., 1991), 104.
3. Paul R. Ehrlich, and Anne H. Ehrlich, *The Population Explosion* (New York, NY: Simon & Schuster, Inc., 1990), 23.
4. Sowell, 215.
5. Ibid., 215.

6. Ibid., 215.
7. Al Gore, *Earth in the Balance: Ecology and the Human Spirit* (New York, NY: Penguin Books USA Inc., 1993), 127.
8. John Tierney, "Betting the Planet," *New York Times Sunday Magazine* (December 2, 1990), 76.
9. Sowell, 211.
10. Ibid., 210.
11. Ibid., 211.
12. Ibid., 212–213.
13. Ibid., 213.
14. James Gwartney, Robert Lawson, and Walter Block, *Economic Freedom of the World: 1975-1995* (Vancouver, BC: The Fraser Institute, 1996), 12.
15. Ibid., 12.
16. Article, "Economic Freedom," *The Economist* (January 13–19, 1996): 22.
17. Ibid., 22.
18. Ibid., 22.
19. Ibid., 22.
20. Ibid., 22.
21. Ibid., 22.
22. Ibid., 23.
23. David Henderson, "Why some Latinos Prosper and Others Don't," *Wall Street Journal*, March 29, 1996. Sec 1.
24. This is due to lack of full overlap in two studies. Almost all data for population and GDP are from 1994. Where 1994 numbers were unavailable, 1992 and 1993 numbers were used.
25. Gwartney, 53.
26. Paul Lewis, "New U.N. Index Measures Wealth as Quality of Life," *New York Times*, May 23, 1993.
27. Mahbub ul-Haq et al., *Human Development Report 1993* (Oxford University Press, 1993), 28.
28. *The World Almanac and Book of Facts, 1995.*
29. China does not appear in either report because its government will not provide enough of the necessary data.

Chapter 19

Problem?

Alan Simmons:

I am very pleased to be here this evening on the thirtieth anniversary of the Pugwash movement. My role this evening is to introduce the debate, to say a little bit about the kind of questions that I hope our panelists will address, to introduce them, and, finally, to make sure that the audience has an opportunity to raise questions to the panelists. I think it should be an interesting and provocative discussion. The topic is particularly appropriate for Canadian Student Pugwash to discuss because the issue does concern the relationship between science and society. I think we know that, in a fundamental way, the progress that has been made in science, particularly in medicine and in food distribution, has led to the kind of population growth which is of concern to us. Falling mortality rates around the world over the past thirty years represent a seemingly good thing yet have also brought about rapid population growth and its controversial effects.

Peter Ross mentioned that approximately one month from today, it is estimated that the world population will reach five billion people. I don't know what five billion means to you; it is a very abstract number. It is five thousand million, and that is a lot of people, especially when you think that in Canada we have about twenty-five million and there seem to be quite a few of us. Twenty-five million is about one-half of 1 percent of the world's population, so that gives you some sense of the relative magnitude that we are talking about.

At the same time, in orienting our discussion tonight, I would encourage us not to think too much about this abstract number. After all, what does this number mean? I think we have to interpret it and put it into the context of other questions. It is a little bit like the issue raised by Thomas Aquinas, the philosopher, when he asked how many angels can dance on the head of a pin.

He did so to point out the futility of certain metaphysical questions. I think the question of how many people the earth can support is a similar abstract, somewhat metaphysical, and maybe in the end, pointless question. Much more to the point is whether the angels dancing on the pin are happy, whether they are healthy, and whether they are wealthy, or at least not poor. I think that is the question which we want to address this evening.

The issue is a complex one and experts in the field of populations disagree with one another on their answers. But rather than trying to come up with global answers, I think that what those of us who are interested in population and development have been doing is to break down the global question into more concrete ones, such as the concrete question of the relationship between environment, pollution, and population growth. Other concrete questions pertain to the relationship between population size and growth, and the health, longevity, and freedom from illness and disability of people around the world. I think they also concern issues of the impact of population growth on income and on access to consumer goods such as cars, televisions, and computers. I raise those kinds of terms somewhat cynically but, in fact, for people in some parts of the world, economic growth is represented by advances in that kind of technology. They also concern issues which are fundamental to the Pugwash movement, that is, the relationship between population growth and world peace, international affairs, and, if we consider all these general issues, we must not forget that we need to put a human face on the matter. Population growth also takes place in families. Children are born, they survive, and they must be supported. We must also take into account the effect that rapid population growth and large families have on the children's parents, particularly on the health and well-being of the mother. Thus, we need to consider the issue of population size and growth not just at the international level but also at the personal level.

These introductory comments should serve to sensitize us to some of the specific issues which I hope our panelists will address this evening. Now, I would like to introduce the panelists. We're extremely fortunate to have two eminent individuals with quite different backgrounds to speak at this debate. One is an economist and the other is a medical doctor. First is Dr. Walter Block, senior economist with the Fraser Institute in Vancouver, and member of the British Columbia Association of Professional Economists. He is a former university teacher and is a very dynamic and energetic writer, his works appearing in a variety of private and public fora. He has published in *Business Week*, the *Financial Post*, and Sterling newspapers, and has written a large number of articles about a variety of economic issues, both national and international.

Our second debater is Dr. George Brown, Vice President and Director of International Programs at the Population Council in New York City. Some of you may know that the Population Council is perhaps the leading international, non-profit organization involved in scientific research and new program development in the population field. Dr. Brown is a specialist in maternal and child health and has spent many of the past twenty years living abroad and developing programs in various parts of the world. I think that he can speak to us with the insight gained through those experiences. For those of you who may not be familiar with the population debate and the issues at hand, I might say just one or two words before turning the floor over to our speakers.

I do not want to guess what our speakers are going to have to say, but I think you might be able to anticipate that in the world at large, there are extremely divergent views on whether increasing population growth is a threat to the welfare of mankind or whether it is simply irrelevant to the major issues which face us at various levels. Dr. Brown is associated with an institution which has taken a very clear stand in this matter. He views the population problem as a fundamental and serious one and, over a number of years, has worked to develop research and to raise public awareness in this area.

Dr. Block comes from an institute which has generally viewed attempts to control, manipulate, interfere with, or generate a movement in areas such as population control, not as a bad thing necessarily, but perhaps as a waste of time. This belief is backed by the idea that there are other, more important things on the agenda. He would likely contend that population dynamics may be much more influenced by the way people organize their life, their society, and the economy, rather than by programs of maternal and child health or by contraception.

With backgrounds such as I have described, I think it is safe to anticipate that our speakers will be addressing somewhat different issues from rather different value perspectives. I will therefore bring my introduction to an end, having sensitized you, I hope, to some of the things to listen for. I would ask the panelists to make an introductory statement of about fifteen minutes each. I will then give them a chance to respond to one another very briefly, followed by a question period. First, Dr. Block.

Walter Block:

The topic that I shall be addressing this evening is "Population Growth: Is It a Problem?" I say no, it is not a problem. I shall defend the position of laissez-faire, as I do on all other questions of public policy. In this case, that means that government should pursue no policy, one way or another, on population size, neither promoting nor opposing population growth. My defense of this will consist of two claims. The first is that all persons have the right to do anything they want, providing they do not initiate violence against other people or violate their rights. Since having children is a peaceful activity, it is justified. My second point is that the arguments in favor of population control are fatally flawed. The case in favor of government policies which have the result of diminishing population is without merit. Let me elaborate.

My first argument against government-sponsored population control is that this is profoundly anti-human. In my view, each person is precious, each person is unique, each person is irreplaceable. If we control population, we might do away with a future Bertrand Russell, Jonas Salk, or Wolfgang Amadeus Mozart. But it does not have to be someone famous who contributes to human society in the magnificent way that these people have done. Those of you that have children of your own know what I mean about each person being precious.

What is the case for government population control? It is stated that we will run out of food if we have too many people. We are treated to various versions of the Malthusian nightmare. This is the idea that population increases in a geometric progression, and food in an arithmetic progression. One famous illustration of this is that if you take a grain of wheat for each square on a checker board and double it as you move through the sixty-four squares; you get an inordinate amount of wheat after a while. In other words, the first box contains one grain of wheat, the second contains two, the third four, on to eight, sixteen, thirty-two, and so on. When you get to the 14th box, you reach 8,192 grains of wheat. The 24th provides 8,388,608 grains of wheat. By the time you finish with the 64 boxes on the checker board, you have 18 quintillion, 446 quadrillion, 744 trillion, 1 billion, 551,615 grains of wheat. This adds up to a billion tons of wheat. We are told that the world population in 1800 was about 0.9 billion, that in 1940 it rose to 2.2 billion, that in 1961 it reached 3 billion, and that this year it will surpass 5 billion. We are scared with this Malthusian nightmare and are treated to scenarios which establish a theoretical level of population on the earth of 20 million times the present population. But this would require the equivalent of a 2,000-story building covering the entire earth and an average of 120 million people per square meter of land. At present growth rates, it would take 890 years to accomplish this. Of course by that time, we would have to have other planets at our disposal. I'm not as frightened by this scenario as some people might be.

The scare tactic is buttressed by the specter of the Ethiopian starvation. But situations like that are not due to overpopulation, rather to improper planning, to centralized government control, to forced collectivization of farms, to price controls on agricultural products, and to Marxist dictatorships. In the Soviet Union, for example, on the 97 percent of the land that is coercively collectivized in large agglomerations of agriculture, only 75 percent of the crops are grown. But the 3 percent of arable land that is under private control accounts for fully 25 percent of the crops. The reason for this is not too hard to fathom. When you unleash individual initiative and the desire to get ahead, miracles occur. Under centralized control, by contrast, people avoid work and slack off. In actual experience, population does not rise geometrically. Food rises faster than population. Right now, we are up to our armpits in grain. There are subsidy wars around the world; people are trying to get rid of these surpluses of grain.

Another point I would like to stress is that, far from being overpopulated, the earth is virtually empty. I traveled to Ottawa from Vancouver today and I looked out the window of the airplane. I am not talking about a flight somewhere up in the Northwest Territories. I'm talking about straight across the southern part of the country, between Ottawa and Vancouver. There is just nobody there! Every once in a rare while you see a few houses or a city, a little teeny, weeny city compared to the vastness the airplane is traversing. I have calculated that if you take all the people in the world, all five billion of us, and put them into little teeny Ontario, each person would have 2,050 square feet. That would provide a family of four with 8,200 square feet, which is roughly the size of a middle-class dwelling with both a front and a back yard. Bear in mind that this calculation takes everyone on the earth and puts them all into the single province of Ontario. The point that I would urge upon you is that the earth is really empty. We might not have room for 2,000-storey buildings, but there is plenty of room to grow.

A particularly objectionable version of Malthusianism is the charge that wages, due to the depressive effect of increasing population, will fall to a subsistence level. This is fallacious, and contrary to all of human history. In order to show this, we shall have to discuss the economics of slavery. My point is that the existence of slavery at any time in recorded history is incompatible with wages being at or below subsistence level. Economically speaking, why would a slave owner be willing to pay a positive price for a slave? This is important because there cannot be a slave system unless someone is willing to pay a positive price for a slave. The only reason the businessman would be willing to pay for a slave, if he were a rational, profit-maximizing slave owner, is if the production of the slave was greater than the cost of caring, feeding, and replacing him. But if the slave's productivity is greater than the cost of such replacement, then his productivity is greater than the subsistence level. And, surely, the productivity of a free person is greater than that of a slave. For one thing, we have incentives. For another thing, there are no costs of guarding and capturing free people in the first place. So from the very existence of slavery we deduce that, in all recorded history, we have never been near subsistence levels except of course for certain unique occurrences such as famines, floods, fires, and so on. We know that slavery existed in biblical times, that it was practiced throughout history, in Roman times, and that it probably existed during the caveman era. This serves as proof that we were never near subsistence level in any of those epochs of human society.

If it is not true that increasing population leads to starvation, then maybe increasing population leads to poverty. In this vein, we are asked to think of the teeming masses in India. Interestingly enough, we are never asked to view the teeming masses in Ottawa, Toronto, Vancouver, New York, Paris, London, or Tokyo. They are not starving or even in poverty. Teeming masses alone are insufficient to explain poverty, and certainly not starvation. The fact of the matter is that there are wealthy countries (per capita income above $5,000, as of 1981) with very high densities (500+ people per square mile). This includes West Germany, Belgium, Holland, and Japan. There are wealthy countries with moderate densities (100–499 people per square mile), such as Kuwait, Denmark, and France. As well there are wealthy countries with very low population density (fewer than 100 people per square mile), such as Sweden, the United States, Nigeria, Canada, Australia, Austria, Libya, and Saudi Arabia. The same applies to the moderately wealthy countries (with per capita income between $1,000 and $5,000). Very densely populated ones include Singapore, Barbados, Jamaica, Taiwan, and Lebanon; moderately densely populated ones include East Germany, Spain, Israel, Greece, Italy, Ireland, Poland, and Costa Rica; and those with very low population density include the Soviet Union, Iran, Iraq, Argentina, South Africa, Brazil, and Mexico. The same sort of distribution exists for the very poor countries (with per capita incomes below $1,000). There are poor, densely populated countries such as El Salvador, India, and Bangladesh; there are poor moderately densely populated countries such as Turkey, Cuba, Malaysia, Morocco, the Philippines, Egypt, Indonesia, China, Uganda, Pakistan, and Burma; and then there are very poor countries with a very low population density such as Colombia, Algeria, Chile, Guyana, Bolivia, Liberia, Congo, Tanzania, Kenya, Afghanistan, and Ethiopia. If we only look at the most densely populated country in the world, which is Macao, with 16,000 people per square kilometer and an annual per capita income of less than $208, then there is support for the view that we should control population. But the second most densely populated country is Monaco. It has the eighth highest standard of living among the 250-odd existing countries. Monaco is a very wealthy country and also very densely populated. So there is no clear pattern which emerges from the statistical record. One cannot deduce poverty from density. The reason we have poverty to the extent that we do is thus not because of density, but I contend that this is because of unwise government interference with the free enterprise system.

There is the argument that we will run out of resources unless government exercises population control. Here I quote from a Sierra Club publication: "Every eight seconds a new citizen is born. He is a disarming little thing but he begins to scream loudly in a voice that can be heard for 70 years. He is screaming for 56,000 gallons of water, 21,000 gallons of gasoline, 10,000 pounds of meat, 28,000 pounds of milk and cream, 9,000 pounds of wheat, etc." These are his lifetime demands on the economy. The point being made is that we will run out of resources if too many people inhabit the planet. Let me divide this argument into renewable and non-renewable resources.

First, let us look at renewable resources. It is utter and obvious nonsense to worry about them because more can always be grown. If a shortage occurs, it is only temporary, and it is due, in all likelihood, to government interference with agricultural markets. If too many trees are cut down on both private property and government holdings, the former decision makers will pay with bankruptcy while the latter will escape such automatic penalties. In the view of Sowell (undated): "It is hard to imagine a more stupid or more dangerous way of making decisions than by putting those decisions in the hands of people who pay no price for being wrong."

Another example is that of the buffalo, which were made almost extinct at a time when the population was much lower than it is today. Yet one can hardly blame overpopulation for their near extinction. On the contrary, the reason the buffalo were made extinct is because the government, in its infinite wisdom, refused to allow the operation of private property rights. Because the buffalo could not legally be fenced in, or farmed, and instead were only hunted, we had what is called the tragedy of the commons. It did not pay for anyone to act in a stewardly way with regard to the buffalo because if a person exercised control, and refrained from mass and senseless killing, he did not benefit from it. Therefore, he might as well kill it even if he only ate its tongue, which is what commonly occurred.

I am sure there are biologists who would point out great differences between the buffalo and the cow, but to an unsophisticated economist there is no distinction to be drawn; they both have horns, and they both moo. But cows were never in any danger of becoming extinct. Why? The reason is that individuals privately own cows and therefore have incentive to take care of them. We have private property rights with regard to cows. Nowadays the whales are becoming extinct. A recent *Star Trek* movie even showed that if we allow the whales to go extinct, ill will betide us in the twenty-fifth century. I think ill will betide us, and a lot sooner than the twenty-fifth century but, again, this has nothing to do with overpopulation. It is because we do not allow private property rights in the oceans. Just as the hunters and gatherers began to farm the land when the population increased beyond tens of thousands of people, so must, we look to the oceans as the world population continues to grow beyond five billion. But until we can farm the oceans, and take advantage or their magnificence and wealth, we will not be able to support the kinds of population that we otherwise could.

What about non-renewable resources? People who worry about them are very unsophisticated. They fail to deal with the fact that whenever we start running out of anything the supply curve shifts to the left and the price rises. When the price of something rises, we automatically bring into play, without any government intervention, the forces of conservation. As the price increases, the amount demanded falls off and we seek substitutes for the commodity. We have very few diamonds, for example, but we are not running out of them because the price rises sufficiently to protect them. Such is also the case with gold, platinum, or whatever may be in very short supply.

It is said that if there is too large a population, there will consequently be too much pollution. This, again, is incorrect. The dust bowl in the United States occurred in a very empty part of the country; there are not too many people around Chernobyl; and oil spills occur in empty oceans. The reason we have pollution has nothing whatsoever to do with overpopulation. Rather, it is because of government mismanagement. Evidence to support my claim can be found by looking at judicial findings from the 1820s and 1830s. In those days, leases were considered under the old laws of nuisance. Nowadays, we would call them the ecology cases. Typical examples included a woman's laundry hanging on the line, being dirtied by cinders blowing in the air, or a farmer's haystack being ignited by sparks blowing from a neighbor's brushfire. These people would go to the government courts and demand an injunction on the basis that, under the law of nuisance, their property rights were being violated. The court would reply that while this is true in theory, public interest is more important than property rights. Thus, public interest, which consisted, in their view, of allowing manufacturing carte blanche was used as an excuse to denigrate private property. By not granting an injunction, the courts were, in effect, telling the manufacturers to go ahead and pollute. Even public-spirited manufacturers did so, because if their competitors all polluted and they did not, they would be at a competitive disadvantage. Thus, business, which could have engaged in non-pollution-intensive structures of production, was instead encouraged to follow pollution-intensive techniques. The great amount of pollution we have today is not the fault of the free enterprise system. Capitalism rests on a system of law which protects private property, but if government destroys this underpinning the market cannot function. In any case, pollution has nothing to do with overpopulation.

True, there is a problem of undersupply of housing, and homelessness in highly populated Ontario. But this is due to rent control. Vancouver also had a problem of high rents and low vacancy rates when rent control was in effect in B. C. But this pernicious legislation was repealed in 1983, and the problem has receded since. As anyone who studies economics will know, demand exceeds supply when prices are held below market levels. That is how housing shortages are created—not because of overpopulation.

My final blow to the argument against population control is that, philosophically, it is illogical. There is a contradictory aspect to it. Let me make this point with regard to the argument for solipsism. This is the view that only, I, the speaker, exist. That may or may not be true. But the problem for arguing for solipsism is that he who does so commits a logical contradiction. One invariably wonders, in such a case who he is talking to and who he is trying to convince.

But the same kind of philosophical law underlies the argument for controlling the population. It proves that even its advocates do not take it seriously. I do not mean to be personal, and I cannot because I do not know what my opponent will be saying, but each of us controls our own lives. If one were seriously worried about overpopulation, the advocate of that view has one option, and that is suicide. The fact that he is still here, talking, arguing, breathing, and living, is contradictory to his stated position. It is, further, hypocritical. It is evidence that he is not convinced by his own arguments. If he were, he has it within his power to lower the population by at least one. And every bit helps according to the mistaken Malthusian philosophy!

In conclusion, I argue the case for laissez faire; that we have a right to bear children, that the government has no right in our bedrooms, and that the arguments for lowering population are without value. Therefore, I counsel against government population controls.

Alan Simmons:
I can see now why they seated the moderator in the middle. Dr. Brown, please.

George Brown:
I am going to address a different problem than the previous speaker. I am going to address the subject of this evening's discussion, which is "Population Growth: Is it a Problem?" Not density, not control, but population growth. My answer to that question is yes, it is a problem. But it is a very different problem in very different societies and it does not help to cast global answers to problems that face different societies and different countries at different times in their history and their social and their economic development. It is really a development issue and I will get back to that in a moment.

I thought I would take a different tactic from Dr. Block, and tell you a little about my background. My involvement in international work began in Haiti, where I worked as a physician in a small town. I very quickly became profoundly concerned with such problems as seeing babies I had treated for kwashiorkor, that is malnutrition, return a few months later with the same problem. I also saw terrible cases of tuberculosis, of infant and child death, and of maternal death. I began to feel that curative medicine could really do very little to alleviate individual suffering because the suffering of humanity seemed to continue onward. I then became interested in public health, and subsequently with family planning. I felt that the problems experienced by mothers from bearing many children at an early age limited the individual choice that we in our society enjoy, of deciding when to have children. These women frequently suffer from ill-health and sometimes death through childbirth. If they choose not to have a child, the women suffer ill-health and sometimes death through abortion. I began feeling that family planning could make an important contribution from a health perspective. As my work continued, both in Haiti and later in North Africa, where I worked for several years, I became involved in the larger questions of how poor countries could address their own development and health problems in a situation of rapid population growth. Again, I emphasize the issue of growth. Getting back to the original comment, I think the real problem is not population size, and in that sense I would agree with my colleague.

The real problem for societies and individuals within those societies is economic and social development. How does one get out of the cycle of poverty, ill-health, illiteracy, poor opportunities for education and, above all, economic poverty? My contention is that because population growth impedes development efforts in terms of health improvement, it is one, but not the only, important underlying problem. It is now very well demonstrated that decreases in family size can decrease infant mortality. There are fifteen million infant deaths per year in the world, most of them in the developing world. Decreases in population growth can improve the availability of health services and other social services that are available to you and me, but which are not available to those who live in Bangladesh, in India, and in sub-Saharan Africa. In that sense, the comparisons with the situations that exist in Canada or the United States are invalid. Population growth is one factor, but not the only factor, which complicates attempts to overcome illiteracy and low education in developing countries, despite enormous investments on the part of these countries to improve their socioeconomic conditions. The governments feel that they have a responsibility to provide education to their citizens just as governments in developed countries do. However, they are faced with very rapid rates of growth, especially of young children coming into the education flow. In many countries, especially in sub-Saharan Africa and Southern Asia, the two broad areas where population growth problems and poverty remain the most dramatic, the governments simply cannot provide an improvement in education. They are providing a lot of education, but it is still at a level that is not overcoming the basic cycle of illiteracy or substandard education that exists. So all the young potential Einsteins who are in those countries—and I assure you there are many who are being born every day—have very little opportunity to receive the education that is required to become an Einstein, or to become simply a doctor, a lawyer, an educator, or a teacher.

Another major issue that is facing many developing countries is a tremendous increase in the population coming into the labor force. This presents a dilemma that is terribly painful, and one which is seen in many parts of the world. An extraordinary increase in young people is entering the workforce with no opportunity for work, thus creating enormous percentages of unemployed or underemployed even among those who have gone through primary or secondary education. Thus, the countries are striving for development by improving educational opportunities, and then are unable to provide jobs for the tremendous numbers of young people that are a direct result of very high fertility fifteen years before.

Food is being provided throughout the world as a means of instigating agricultural self-sufficiency, yet there remains tremendous disparity. Some developing countries have done extraordinarily well in increasing per capita agricultural produce, but one important part of the world, sub-Saharan Africa, has seen a decrease in per capita agricultural output over roughly the past ten years.

It is fine to say that Canada, the United States, Australia, and Argentina produce huge surpluses of wheat and grain, which is true, but it is another matter to see whether that grain can get to people in parts of the world that do not produce enough. It is even worse, in my opinion, that this situation creates a dependency among those countries on handouts from the wonderful granaries and wheat fields of Alberta and Saskatchewan. This situation exists and is recognized by Canada, the United States, and other grain-producing nations. There are indeed major efforts being made to improve the situation, such as the development of miracle grains, miracle rice, and miracle wheat.

But we must ask ourselves why this problem exists in the first place. There are many reasons, one of them being the fact that there is disparity or imbalance in terms of population growth and agricultural output. But this cannot be the only issue since there are other parts of the world which also have rapid growth, such as the Punjab in India, but which have turned around and made a miraculous effort in improving their own agricultural output. It is true, as my colleague says, that there are many examples that can be used to support both sides of the argument, but they do not apply to all countries. Just as government leaders have to calculate the rate of annual agricultural output, they also have to balance it against the number of mouths to be fed. To ignore that would be to neglect their duty. Therefore, population growth is an issue.

Another important reason for population being a problem in terms of development is the tremendous rate of urban growth in the past thirty years, especially in the developing world. Urban growth by itself is fine. For better or worse, I happen to live in one of the enormous conglomerations of the world, namely, New York City. Formerly the largest city, its population is about to be surpassed by Mexico City, Sao Paulo, and Bombay, to name a few. It is neither the size nor the density that is important, when human beings live close to each other in extremely large cities. Rather, it is the ability of governments to structure the services that provide such basic elements for people as food, transportation, waste disposal, and water supply. It is fine to have many people in such cities as Toronto, where the subways are clean, it is easy to move around, there are convenient ways to get out in the countryside, reliable water is available, and garbage disposal is provided. Unfortunately, this is not the case for many of the developing countries and, by the way, parts of New York as well. We have enough trouble coping with those kinds of basic urban services in our own cities, despite all our wealth. I have visited many of the cities in the developing world and, although there are fine parts of them where the tourists go and where amenities exist, there are unfortunately huge, huge areas of urban slums where there is nothing approaching services for basic human welfare.

Some countries, however, have successfully improved their development standing despite fairly rapid population growth, and on that point I would agree with my colleague. However, most of those who have achieved economic success have, at the same time or roughly parallel to it, undergone decreases in the rate of population growth. I am thinking of Brazil, Thailand, South Korea, Taiwan, and others, to say nothing of the very special city-states such as Singapore and Hong Kong. There are many reasons for the achievement of development goals and there are many reasons for failure. I argue that those countries which have, at least in part, been able to cope with their development problems, in spite of rapid growth, still have major problems because there are huge disparities within the countries. The south of Brazil, for example, is as developed as Canada, whereas the northeast is still very poor and struggling.

My major concern focuses on the parts of the world where I have been most involved over the past five years; that is, South Asia, including India, Pakistan, Bangladesh, and Nepal, and sub-Saharan Africa. In those parts of the world, population growth remains high despite falling mortality, and major efforts by those countries and regions to improve their economic and social well-being have been fraught with problems. The biggest problem right now in sub-Saharan Africa and some other parts of the developing world, I would certainly agree, is not population but the economic debt problem. I will leave my economist colleague to deal with that issue. I certainly would not for a moment say that population is the only issue there, but it does play a role. Zambia, for instance, a country in which I have been working recently, is now confronted with enormous debt problems and a population growth of something over 3 percent per year. Therefore, it has to improve its per capita income by 3 percent just to remain stable. Unfortunately, it is unable to do that and the per capita wealth of that country is falling dramatically. It is getting poorer partly because of population growth but, more acutely right now, because of economic decline.

I think that each country needs to analyze its own population situation. Canada has a different population problem from many countries since it is increasing its population through immigration. It is fine to say that Canada has open spaces, but I would not like to be one of those who is given Dr. Block's corner of Ontario up in the tundra. Other countries also need to increase their population growth. Several European countries are very concerned about declining growth and so they should be. Each country needs to analyze its own population situation, growth distribution, and migration in light of its own circumstances and the needs of its own people. Finally, the real crux of the matter is the choice of individual families to decide how many children they want. In order to do that, they must have access to full availability of contraceptive information and services through family planning programs which can be private or public, which should be voluntary but, most importantly, which should be available throughout the world just as they are available to all Canadians.

Alan Simmons:

That ends the more formal part of our debate. I am now going to ask our speakers to sit reasonably close together and to look one another in the eye from time to time. I would like to begin by giving them the chance to respond to one another. I would like to ask Dr. Block if he would like to raise any points to sharpen the debate between himself and Dr. Brown.

Walter Block:

First of all, I don't see the difference or distinction between growth and density. Surely, what we fear growing toward is high densities. Second, Dr. Brown says the real problem is not economic or population growth or density, but economic development. I would certainly agree with that. He says that population growth impedes development, but this is shown to be false by all those very wealthy countries that have had high growth rates in the past and consequently, have high density rates now. I am referring, for example, to Japan, Monaco, and Hong Kong.

With regard to government education, I do not see why we should always resort to and look to the government for everything. Government is an intrinsically coercive institution. It is the only institution in society that obtains its revenue through coercive means, through the tax system. Therefore, what we should do is to minimize our reliance on government, not maximize it.

I also take issue with his reference to many of the countries he mentions as developing countries. Who says they are developing? A non-pejorative reference might be to say they are underdeveloped and leave it open as to whether they are developing or not. They will develop all right if they follow the pattern that already developed countries have followed, namely, a great reliance on free enterprise institutions.

To say that population size or density or growth has anything to do with unemployment is false. There are very low unemployment rates in very highly densely populated places, such as Hong Kong and Japan. And there are very high unemployment rates in countries like Canada which have an infinitesimally small density. Dr. Brown talks about the Western agricultural groups producing grain while Ethiopia does not. But the question is why not? Why is it that we are very productive here and not there? I think a tremendous part of the problem is not population size or growth or density, but rather the relative economic freedom of the different nations.

If we are worried about underdeveloped countries, and I think we should be, then we should pursue one policy that has not been mentioned, so far tonight, and that is free trade. Yet we find that, hypocritically, the same people who are talking about underdeveloped countries and are shedding all sorts of crocodile tears about them, like Pierre Trudeau, are the most vociferous opponents of free trade. Opening international markets is a way of integrating the economies of the Third World with our own. I think we do them a vast disservice by not allowing them to sell us shirts at $3 a piece or shoes at $5 a pair. A large part of the explanation for this sorry state of affairs is that we as a nation are unduly enthralled by industries, such as the textile and the shoe manufacturers, that are highly inefficient and that are hiding behind a wall of protective tariffs. If we want to help the underdeveloped countries, what we should do is end foreign aid right away and integrate them and their economies through foreign trade. Instead, we give them welfare which mainly goes toward the three Ms—monuments, Mercedes, and machine guns—which better enables the Third World dictators, who are responsible for the poverty of their peoples in the first place, to gain further control over them. What we need is less dole, fewer attempts to take away these peoples' incentives, and more trade and honest commerce with them.

Dr. Brown also talks about government leaders worrying about feeding the population. How can the government leaders care, since they choose to follow policies that make it impossible for their people to get fed in the first place? I think the fewer government dictators we have, and the less power they have, the more likely it is that people will be fed. The first few countries that burst out of economic underdevelopment were Great Britain and the United States, and they did it with minimal help from government. I think if we are serious about the Third World wrestling its way out of the despicable poverty it presently wallows in, we have to encourage them to follow the same principles.

Dr. Brown talks about New York City, which is an economic basket case. But he fails to acknowledge that it is that way because of too much government. New York has a great amount of people on the welfare dole. The South Bronx has problems due mainly to rent control. New York City has had rent control since 1941 because of the wartime emergency measures of World War II. But here it is forty-six years later and their wartime emergency measures are still in place! That is why the housing stock is disappearing. The economic debt is not the reason that the underdeveloped countries are in trouble. Economic debt means that the West lent money not to commercial people in those Third World countries but to the central planners who cannot plan their way out of a paper bag.

Dr. Brown also claims that each country has to decide on population size. I do not think so. I agree with him when he says that each person or each family has to decide, and I certainly would agree that there should be information on birth control methods. But surely the voluntary private sector could be relied upon in this regard. There are people who object to this material on religious grounds, and it seems to add insult to injury to force them to support things that they do not believe. Catholics, for example, oppose birth control; why should we take their money from them at the point of a gun through the tax system and use if for a project they regard as immoral? I think this is a perverse use of government. It is on these kinds of issues that I would disagree with my learned colleague.

George Brown:

I cannot quite understand why my colleague does not understand the difference between growth and density. They are two very distinct things. There are poor countries that are extraordinarily empty of people. One country I know very well is Mali, whose density, I would guess, is less than Canada's. It certainly has empty spaces in the Sahara desert just as Canada has in the Great North. Its problem is not the number of people per square meter but the issue of population growth, economic growth, and, above all, human capital growth. Those elements all need to be figured into a country bereft of almost any national resources other than limited, relatively poor agricultural territory and the aspirations of the people. I am sad to have to come up to the country of my birth that I love, to hear "Reaganomics" all over again. And I think that since I am not an economist, I am not really prepared to argue "Reaganomics."

With regard to the development field, I would agree that the emphasis on privatization, on getting agricultural prices right, and on getting heavy-handed government control removed, has been a positive thing, and I applaud it. On the other hand, what I seem to hear sounds like what it must have been like in the summer of 1929 under the regime of President Hoover. I just cannot believe that a total laissez-faire policy is even remotely imaginable. But perhaps I have got just too limited a brain for my learned colleague and his "Reaganomics," which I hope will only last another year.

He seems to have not really referred to the health issue, and I feel very strongly about that. Perhaps my sensitivity to the issue comes from my long experience in developing countries. I have seen too many babies and mothers die, and too many problems of abortion which could have been at least in part alleviated by the free availability of family planning. I certainly agree with Dr. Block in terms of the need for voluntary availability of family planning, although I think the crux of the matter there is that governments of nations do have responsibilities. Why do we have a Pugwash movement to bring together the Americans and the Russians? We have governments. We have a United Nations. My opponent would seem to want to do away with these institutions. Perhaps that would be a good thing, I don't know. I just think it is so unrealistic, so utopian, and so unfeasible that it is setting up the kind of debate that does not grapple with the real problems, as I have lived them and perceived them. But let me close with one other comment. I do agree that the Malthusian idea of population explosion with people dropping off the globe and vast numbers of skyscrapers being erected is all absolute nonsense. It is something that has been out of the debate in the international development field for at least fifteen to twenty years, except for some radical fringe on the other side. We are talking about a very complex world in which issues of agriculture, education, health, and poverty must all figure in with population to create an equation with which we all seek to improve humankind in the developing countries.

Alan Simmons:
I think the debate is picking up focus and momentum. I would like to turn, then, to the audience and to questions that people would like to raise to either panelist or to the panel as a whole.

QUESTION PERIOD

Audience (question 1):
I agree with many of the points made by Dr. Block, however, I have a feeling that, in some senses, he is focusing more upon the developed nations and Dr. Brown is focusing more upon the underdeveloped nations' problems. Dr. Block, how would you respond to the very real problem of population increase within underdeveloped countries, which is growing at a higher rate than the actual productivity?

Dr. Block:
My short answer would be to recommend you to the books of Peter Bauer, one of the Fraser Institute authors, who is, as far as I am concerned, the world's foremost authority on this question. He shows that underdeveloped countries are impoverished because they follow economic policies that lead to economic underdevelopment. These nations concentrate on government bureaucrats deciding things; they denigrate both the profit and loss institutions, and private property rights, and they rely on foreign aid, which politicizes them. The foreign aid that we give to them is not really much by our standards, but from their point of view it is very significant. What happens is that instead of the young people in those countries concentrating on matters of industry and commerce, which is the

way the United States and Great Britain worked their way out of underdevelopment, they start concentrating on political careers in order to gain control of this largesse. This moves them in an improper way.

I disagree with your characterization of me being unconcerned with underdeveloped countries; I certainly mentioned them. I was concerned with tracing the question of whether high or rising population, or rising population density, is correlated with all these evils. Take poor health, for example. I am against little babies dying, and the misery of sickness; but I do not think that there is a great correlation with population growth or density. There are countries that have high population growth and density, but do not have high percentages of unhealthy citizens, just as there are countries that have low population growth and density and do have high percentages of unhealthy citizens. The reason, on the contrary, has to do with reliance on the market place. Those countries that place a great reliance on the market place are the ones that do not have babies dropping off like flies, nor do they have dire poverty. I think that the way for the Third World countries to rise up to the First World level is to follow the same paths that we followed to get to those levels; namely, a reliance on the market. And the same applies to Hong Kong, Singapore, South Korea, and Taiwan. If this be "Reaganomics," so be it. Why should we be put off by a term? If Reagan is following these policies—which I happen to believe he is not—but if he is, there is nothing wrong with that.

Audience (question 2):

I wish to address Dr. Block with regard to his premise that it is the right of each individual to do as he pleases. I think, with the way the world is going these days that we need to consider whether that is not a very ethnocentric attitude. The deification of the individual over the society is something that has resulted in the rather prosperous West, and perhaps what we should instead be looking at is what will benefit society as a whole, even though it may not be in the direct best interest of the individual.

Walter Block:

First of all, I did not say that people have a right to do whatever they want. I said that people have a right to do whatever they want as long as they do not violate the rights of other people. I fail to see how my having a child can ipso facto violate someone else's rights. Conceivably, it could if high density or high growth rates had the nefarious effects outlined by Dr. Brown, but they do not. With regard to deification of the individual over society, I do not ascribe to that view. I think society is a fine thing, but I am in favor of voluntary interactions of individuals because individuals are the only people that can act. Societies cannot act apart from the actions of the individuals that comprise them. I am a methodological individualist in the sense that I do not see that there is a group mind. There is no group mind that I am a part of. I am an individual. I choose right or wrong, good or bad, and so do we all. Only individuals can act and I think only individuals can have rights. Societies cannot have any rights that are not made up of the rights of their constituent elements, namely, individual human people.

George Brown:

I would add a comment to that. I think that the societies do make a limitation on our behavior. We are only allowed to marry one spouse at a time, we have to obey red lights or risk certain penalties, and we have to drink chlorinated water. There are a number of things that we as groups of human beings agree to do, sometimes for the betterment of ourselves as individuals, and sometimes for the betterment of society. I think this question really does relate to the question of whether it is possible that individual behavior and reproduction over time can create a problem for the state. I think it can, although there are many other variables. One of the problems I have in this debate is what I see as a fairly simple approach to extraordinarily complex issues. It is, indeed, a fact that in China, the most populous country in the world, the government has felt that the size of population, now rapidly approaching one billion people, is growing too fast for its collective needs, and is now undertaking policies that I, for one, am very uncomfortable with. I hope that in other parts of the world there will be an earlier recognition of the problem. Information and services can enable individual couples to make their own decisions and partake in voluntary family planning in order to ensure that the kind of problem that is now presented or is perceived by the government of China will not occur elsewhere.

Audience (question 3):

The topic we are discussing is "Population Growth: Is It a Problem?" We heard Dr. Brown tell us of his personal experience in Third World countries. I think, because of his personal experience, he has an understanding at the emotional level of the kind of suffering that goes on in those countries. We also heard from Dr. Block who has suggested some long-term solutions to the problem. My question to the panelists is: Does geographical distance exculpate us from responsibility? It we have hungry children in this country, we feed them, we educate them, we train them, and we provide them with the tools to become self-sufficient. In many underdeveloped countries, the children don't have opportunities, education, or food. Under those circumstances, I think we have a responsibility to do what we can to ensure that they have the opportunities to pull themselves out. I am not disagreeing with the economic policies of Dr. Block, but I am suggesting that we should be sensitive to the needs of these people. We must not just look at the long-term solutions, but recognize that right now people are suffering and dying.

Walter Block:

I agree heartily that we have a responsibility to help these people, especially the starving babies, who are innocent. However, in my opinion, foreign aid does the opposite of helping them. I think that government transfers of funds simply allow the Third World dictators to further cement their control over these helpless people. I think we do have a responsibility and it does not I diminish one bit with distance. It is a philosophical point, not a geographical point, and I urge you to propose and propound and defend the market system, which is the last best help of these poor people.

Audience (question 4):

One might say that pollution is a result of consumerism and therefore, if there are more consumers, will we not have more pollution? As Third World countries become developed, will they not also want the lifestyle that we have and therefore become consumers? Also, I have been wondering about the infringement on the wilderness. We have a lot of land which we could use to grow grain on, but can we justify doing that? Should we not save the wilderness for what it is? And if one is going to have a world full of private ownership, who would own the wilderness and what use would it be to them if it is just for preservation? What benefits could they have?

Walter Block:

The reason we have pollution is not because of the population. We had lots of pollution in the nineteenth century with many fewer people. The reason we have water pollution is because of improper government laws. Government has fallen down on its duty to set the rules of the game. Regarding ownership of the wilderness, in a free society the market could also supply another good or service, that being parks. In addition to people who enjoy visiting Disney World and city parks, there are backpackers and Sierra Club types of people who like to have wilderness preserved. The Sierra Club, for example, owns vast acreages of wilderness areas and serves as a perfect example of the free enterprise system at work. The Sierra Club is a private institution that went out into the market on a cooperative basis, which is not incompatible with the market system, and brought up large acreages which it holds as wilderness preserves. I think that it is a much more viable system than allowing political pull between Ottawa and the provinces to determine whether we are going to have parks or not. I think that the Sierra Club and entrepreneurs who would supply a service to consumers are much more likely to be relied upon to preserve wilderness than is the government.

It is important for the individual families to determine these things. If there are no tractors, and the people all lack strong backs and strong arms, who will do the planting? The point, though, is why is it that there are not any tractors in some parts of underdeveloped countries? Why is it that multinational corporations are not in those areas with tractors and better technology? One reason is that they have been badly burned in the past when they invested capital in the Third World countries. They were subjected to nationalization and the resulting expropriation of their property, leaving them with very little at all. So, once burned, they are twice shy. It was like killing the goose that laid the golden eggs. These Third World tin pot dictators get up on their hind legs and expropriate the private property of these multinational corporations, and then wonder why the multinational corporations are not forthcoming with more, not just with capital, but with learning, knowledge, and training for the indigenous labor population. The reason is that they have been scared off by a violation of property rights. I think the best policy for the Third World countries is to leave off nationalization and expropriation.

Audience (question 5):

I feel Dr. Brown represents the people that are dealing with the situation that exists now and that Dr. Block is representing the people with the long-term solutions. We need both, and I think we need cooperation between the two. I do not like to see the two opposing each other as much as they are.

George Brown:

I think that is a very good point. I tried to say in my earlier remarks that, despite my crack about "Reaganomics," I think that those who have been concerned and working with seeking cures to problems of poverty in the world have indeed benefited over the past several years with the notion that more emphasis be placed on private resources, private enterprise, and the reduction of government control. As far as I can tell, not being an economist, I do feel that this change has been a positive element. In family planning, there is certainly strong support of private, non-government, voluntary, and other institutions becoming involved in communities, seeking their own ways to approach the issue of provision of family planning services. My problem comes in trying to envision the world that Dr. Block describes, and that is what we have been talking about.

Walter Block:

I think that, paradoxically, we cooperate through the competitive system both economically and in this debate. We have different views. In a sense, we are competing, but we are also cooperating. Hopefully out of the synthesis and antithesis something better will emerge. One of the sharp disagreements, of course, is Dr. Brown saying that my views are utopian, unfeasible, and politically impossible. I would stipulate that that is true; they are. But at one time the view that Earth is round was inconceivable and unbelievable. I would say that even though my views are politically unfeasible right now, if people were to change their minds, my ideas would be less politically unfeasible. I think the truth is on the side of the points that I have been making.

Audience (question 6):

I would like to ask Dr. Brown if he felt that since many of the world's problems are immediate ones, how the problem of population growth, as an immediate problem, can be solved by Dr. Block's approach of laissez faire.

George Brown:

I do not think that population control can resolve immediate Third World problems except, perhaps, in one specific area, simply because population must be viewed in a long-term perspective. Those infants born this year will be seeking jobs fifteen and twenty years from now. The problem of old age, for instance, is something that has a long-term perspective. That is one of the problems in population; one must look at long-term issues while facing short-term crises. The one area where I do think that well-thought-out population policies can contribute in the short run is in the alleviation of high infant and maternity mortality. That, indeed, is an area that I have been involved in over the past years.

Audience (question 7):

Dr. Brown, I got the impression from your concluding remarks that you think each underdeveloped country should look at its own resource and population problems and solve them themselves. I was curious to hear that from you. I would have expected a more global approach. Could you perhaps extrapolate on that matter or elaborate why you think it should be a national policy as opposed to a world concern.

George Brown:

Whereas I do agree that the world needs to look at the issue of population policies, the questions surrounding global population growth are so diverse and so varied that certain difficulties arise. Solutions to problems in Ontario, for example, are not of much help when dealing with the government of India, for example, which I must say is the largest democratically elected country in the world.

Walter Block:

If people in the field of overpopulation had a philosophically consistent point of view, they would not say that each country should decide for itself. They would have a theory as to how each country should decide. In contrast, the theory of laissez-faire does not say each country should decide for itself whether to have socialism or capitalism. Rather, it proposes a theory that can be applied to all countries.

Alan Simmons:

I would like to assure you that there will be no final word in this debate. We have had the benefit of the time, the energy, and the very serious thought and reflection of two people with a range of quite contrasting views and experience, but between whom some overlap started to show up throughout the debate. Being the moderator turned out, in the end, to be somewhat less dangerous than I had feared at one point. I would like to thank both of our debaters for having agreed to be with us.

Alan Simmons is the director of the Centre for Research on Latin America and the Caribbean as well as an associate professor in the Department of Sociology at York University.

Walter Block is senior economist at the Fraser Institute in Vancouver, and Director of its center for the study of economics and religion.

George Brown is vice president of the Population Council and director of the International Program, New York. He is a physician and public health specialist, having obtained his M.D. at the University of Toronto and his M.P.H. at Harvard University.

Bibliography

Agence France-Presse. 2002. "Agency Puts Hunger No. 1 on List of World's Top Health Risks," October 31. *The New York Times*. https://www.nytimes.com/2002/10/31/world/agency-puts-hunger-no.1-on-list-of-world-s-top-health-risks.html (accessed November 4, 2002). www.nytimes.com.

American Nuclear Society, University of Texas Chapter. "The Shoreham Plant..." http://www.me.utexas.edu/~ans/Anti/shoreham.html (last accessed January 19, 2006).

Anderson, Martin. 4 January 1989. *The Christian Science Monitor*, p. 19, reprinted in Block, 1990a.

Anderson, Terry L. and Peter J. Hill. 1983. "Privatizing the Commons: An Improvement?" *Southern Economic Journal*, Vol. 50, No. 2, pp. 438–50.

Anderson, Terry L., and Leal, Donald R. 1991. *Free Market Environmentalism.* Pacific Research Institute for Public Policy: San Francisco, CA, pp. 1–24.

Anderson, Terry L. 2000. *Political Environmentalism: Going Behind the Green Curtain.* Contributors: Anderson, Terry L.; Adler, Jonathon H.; Yandle, Bruce; Lueck, Dean; Swope, Kurtis J., Benjamin, Daniel K., and Anderson, Terry L.; Brubaker, Elizabeth; Gerard, David; Stratman, Thomas; and Morriss, Andrew P. Hoover Institutional Press. Stanford, CA. Stanford University. Copyright 2000, pp. x–xv, 2, 32–47, and 245.

Anderson, Terry L. and Donald R. Leal. 2001. *Free Market Environmentalism.* New York: Palgrave.

Anderson, William, Walter E. Block, W., Tom DiLorenzo, Ilana Mercer, Leon Snyman, Christopher Westley. 2001. "The Microsoft Corporation in Collision with Antitrust Law," *The Journal of Social, Political and Economic Studies*, Vol. 26, No. 1, pp. 287–302.

Anderson, William L. 2002. "ANWR and Private Property." http://www.mises.org/fullstory.asp?control=941.com

Armentano, Dominick T. 1999. *Antitrust: The Case for Repeal*, revised 2nd ed. Auburn, AL: Mises Institute.

Baden, John and Richard Stroup. 1989. "Natural Resource Scarcity, Entrepreneurship, and the Political Economy of Hope," in *Environmentalists versus the Economy*, edited by Walter Block. Vancouver: The Fraser Institute.

Baden, John and Stroup, Richard L., eds. 1981. *Bureaucracy vs. Environment*. Authors: Baden, John and Stroup, Richard L.; Fort, Rodney D. and Baden, John; Anderson, Terry L. and Hill, Peter J.; Orr, Lloyd D.; Habict, Ernest R., Jr.; Libecap, Gary D. and Johnson, Ronald N.; Shanks, Bernard; Kremp, Sabine; Lanner, Ronald M.; Dowdle, Barney; Hyde, William; and Johnson, Bruce. The University of Michigan.

Baird, Charles. 1980. *Rent Control: The Perennial Folly*. Washington, DC: The Cato Institute.

Barnett II, William. 1989. "Subjective Cost Revisited," *The Review of Austrian Economics*, Vol. 3, pp. 137–138.

Benjamin, Daniel K. 1993. "Risky Business: Rational Ignorance in Assessing Environmental Hazards," in *Meiners, Roger and Yandle, Bruce's Taking the Environment Seriously*. Lanham, MD: Rowman and Littlefield Publishers, Inc, p. 224.

Benson, Bruce L. 2004. "Federal Immunity from Toxic Substances," PERC Reports. Property & Environment Research Center. 4 November http://www.perc.org/publications/percreports/march2001/government.php.

Berkes, Fikret, et al. March 17, 2006. "Globalization, Roving Bandits, and Marine Resources," *Science*, Vol. 311, pp. 1557–1558.

Birchall, J., Roberts, D. and Taylor, P., eds. 2005. "Federal Agency 'Slow' to Accept Business Help," available at: www.ft.com (accessed April 29, 2007).

Block, Walter. 1976. "The Ticket Scalper," in *Defending the Undefendable*. New York: Fleet Press.

Block, Walter. 1979. "Free Market Transportation: Denationalizing the Roads," *Journal of Libertarian Studies: An Interdisciplinary Review*, Vol. III, No. 2, summer, pp. 209–238; http://www.mises.org/journals/jls/3_2/3_2_7.pdf; anthologized in *The Libertarian Reader*, edited by Tibor R. Machan. Totowa, NJ: Rowman & Littlefield, 1982, pp. 164–183.

Block, Walter. 1989. "Population Growth: Is It a Problem?" *Resolving Global Problems into the 21st Century: How Can Science Help Proceedings of the Fourth National Conference of Canadian Pugwash*. Ottawa: CSP Publications, pp. 30–61.

Block, Walter. 1990. "Earning Happiness Through Homesteading Unowned Land: A Comment on 'Buying Misery with Federal Land' by Richard Stroup," *Journal of Social Political and Economic Studies*, Vol. 15, No. 2, Summer, pp. 237–253.

Block, Walter, ed. 1990a. *Economics and the Environment A Reconciliation*. Vancouver: The Fraser Institute.

Block, Walter. 1990b. "Earning Happiness Through Homesteading Unowned Land: A comment on 'Buying Misery with Federal Land' by Richard Stroup," *Journal of Social Political and Economic Studies*, Vol. 15, No. 2 (summer), 237–253.

Block, Walter. 1994. "Total Repeal of Anti-trust Legislation: A Critique of Bork, Brozen and Posner," *Review of Austrian Economics*, Vol. 8, No. 1, pp. 35–70.

Block, Walter. 1998. "Roads, Bridges, Sunlight and Private Property: Reply to Gordon Tullock," *Journal des Economistes et des Etudes Humaines*, Vol. 8, No. 2/3, June–September, pp. 315–326; http://141.164.133.3/faculty/Block/Blockarticles/roads2_vol8.htm.

Block, Walter. 1998. "Environmentalism and Economic Freedom: The Case for Private Property Rights," *Journal of Business Ethics*, Vol. 17, p. 1887.

Block, Walter. 2002. "A Critique of the Legal and Philosophical Case for Rent Control," *Journal of Business Ethics,* Vol. 40, pp. 75–90; http://www.mises.org/etexts/rentcontrol.pdf.

Block, Walter. 2006. "Katrina: Private Enterprise, the Dead Hand of the Past, and Weather Socialism; An Analysis in Economic Geography," *Ethics, Place and Environment: A Journal of Philosophy & Geography*, Vol. 9, No. 2, pp. 231–41.

Block, Walter and Matthew Block. 1996. "Roads, Bridges, Sunlight and Private Property Rights," *Journal Des Economistes Et Des Etudes Humaines*, Vol. VII, No. 2/3, June–September, pp. 351–362; http://141.164.133.3/faculty/Block/Blockarticles/roads1_vol7.htm.

Block, Walter and Matthew Block. 2005. "Private Parks and Walkways Under Free Enterprise," *Ethics, Place and Environment: A Journal of Philosophy & Geography*, Vol. 8, No. 2, June, pp. 201–208.

Block, Walter, Joseph Horton and Ethan Shorter. 1998. "Rent Control: An Economic Abomination," *International Journal of Value Based Management*, Vol. 11, No. 3, pp. 253–263.

Block, Walter, and Edgar Olsen, eds. 1981. *Rent Control: Myths and Realities*. Vancouver, The Fraser Institute.

Block, Walter, and L.H. Rockwell, Jr. 2007. "Katrina and the Future of New Orleans," *Telos*, Vol. 139, pp. 170–185.

Block, Walter and Roy Whitehead. 1999. "The Unintended Consequences of Environmental Justice," *Forensic Science International*, Vol. 100, p. 57.

Boettke, P.J. (1991), "The Austrian Critique and the Demise of Socialism: The Soviet Case," in *Austrian Economics: Perspectives on the Past and Prospects for the Future*, edited by R.M. Ebeling. Hillsdale College Press, Hillsdale, MI, Vol. 17, pp. 181–232.

Bookchin, Murray. 1970. "Toward and Ecological Solution," *Ramparts* (May).

Bookchin, Murray. 1989. "Death of a Small Planet," *The Progressive* (August).

Boudreaux, Don J. and Thomas J. DiLorenzo. 1992. "The Protectionist Roots of Antitrust," *Review of Austrian Economics*, Vol. 6, No. 2, pp. 81–96.

Bramwell, Anna. 1989. *Ecology in the 20th Century*. New Haven: Yale University Press.

Brown, Lester R. November/December 1991. "The Environmental Crisis: A Humanist Call for Action," in *The Humanist*, pp. 26–30.

Brownstein, Barry P. April 17, 1984. "The Price-Anderson Act: Is It Consistent with a Sound Energy Policy?" *Cato Policy Analysis*, No. 36. http://www.cato.org/pubs/pas/pa036.html (last accessed January 19, 2006).

Buchanan, James M. 1969. *Cost and Choice: An Inquiry in Economic Theory*. University of Chicago Press.

Buchanan, James M. and George F. Trilby, eds. 1981. *L.S.E. Essays on Cost*. New York Univ. Press.

Buchanan, James M. and Gordon Tullock. 1971. *The Calculus of Consent*. Ann Arbor: University of Michigan Press.

Byrd, Deborah. 1998. "Quick: Define Organic in 10 Days of Less. (Originated from Knight Ridder Newspapers). Knight Ridder/Tribune News Service. April 20, 1998. P420K1497. Gale Group, p. 1.

Caplan, Bryan. 2007. "Special-interest Secret," *Wall Street Journal*, available at: http://online.wsj.com/article_email/SB117893365787300771-lMyQjAxMDE3NzE4MjkxMzIzWj.html (accessed May 12).

Childs, Roy A. Jr. 1977. "The Invisible Hand Strikes Back," *The Journal of Libertarian Studies*, Vol. 1, No. 1, pp. 23–34.

Clanton, Adam. Winter 2006. "Enforcing Individual Rights in an Industrial World: Legal Rules and Economic Consequences," *The Georgetown Journal of Law & Public Policy*, Vol. 4, No. 1.

Coase, Ronald H. 1960. "The Problem of Social Cost," *Journal of Law and Economics*, Vol. 3, pp. 1–44.

Commoner, Barry. 1990. *Making Peace with the Planet*. New York: Pantheon Books.

Conquest, Robert. 1986. *The Harvest of Sorrow*. New York: Oxford University Press.

Conquest, Robert. 1990. *The Great Terror Edmonton*. Alberta: Edmonton University Press.

Cordato, Roy E. 1989. "Subjective Value, Time Passage, and the Economics of Harmful Effects," *Hamline Law Review*, Vol. 12, p. 229.

Cordato, Roy. 2004. "Toward an Austrian Theory of Environmental Economics," *The Quarterly Journal of Austrian Economics*, Vol. 7, No. 1, Spring, pp. 3–16.

Cornejo, Fernandez and McBride, William. 2000. "Genetically Engineered Crops for Pest Management in U.S. Agriculture: Farm-Level Effects." Resource Economics Division, Economic Research Service, U.S. Department of Agriculture. Agricultural Economic Report No. 786. April. 1800 M Street, NW. Washington, DC 20036-5831.

De Alessi, Louis. 1998. "Private Property Rights as the Basis for Free Market Environmentalism," in *Who Owns the Environment?*, edited by Peter J. Hill and Roger E. Meiners, Rowman & Littlefield Publishers Inc.

DeMont, John. 2003. "The Empty Seas." *Maclean's*. Toronto, Vol. 116, No. 44, November 3, p. 55.

Diamond, A.M. Jr. 1987. "The Intergenerational Invisible Hand: A Comment on Sartorius's Government Regulation and Intergenerational Justice," *The Journal of Libertarian Studies*, Vol. 8, No. 2, pp. 269–274.

DiLorenzo, Thomas J. 1990A. "*Does Capitalism Cause Pollution*?," St. Louis, Washington University: Center for the Study of American Business, Contemporary Issues Series 38.

DiLorenzo, Thomas J. 1990B. "The Subjectivist Roots of James Buchanan's Economics," *The Review of Austrian Economics*, Vol. 4, p. 180.

Dininny, Shannon. 2004. "State Looks into how to Enforce Hanford Initiative." Seattle Post-Intelligencer. 4 November <http://seattlepi.nwsource.com/local/198151_hanford04.html>.

Ebeling, Richard M. 1993. "Economic Calculation Under Socialism: Ludwig von Mises and His Predecessors," in *The Meaning of Ludwig von Mises*, edited by Jeffrey Herbener. Norwell, MA: Kluwer Academic Press, pp. 56–101.

Economist. 2000. "Growing Pains." Economist.com. May 23. www.economist.com/displayStory.cfm?Story_ID=S%26%298H%29Q17%21%0A&ppv=1 (accessed November 4, 2002).

Economist. 2002A. "Better Dead than GM-fed?" Economist.com. www.economist.com/science/displayStory.cfm?story_id=1337197 (accessed October 22, 2002). https://www.economist.com/science-and-technology/2002/09/19/better-dead-than-gm-fed (accessed March 25, 2020).

Economist. 2002B. "Seeds of Uncertainty." Economist.com Feb 7th. www.economist.com/displayStory.cfm?Story_ID=S%26%2BH0%29Q%21%21%0A&ppv=1 (accessed November 4, 2002).

Ehrlich, Paul R. and Anne H. Ehrlich. 1981. *Extinction: The Causes and Consequences of the Disappearance of Species*. New York: Random House.

Ehrlich, Paul R. and Anne H. Ehrlich. 1990. *The Population Explosion*. New York: Simon & Schuster.

Ehrlich, Paul R. 1968. *The Population Bomb*. New York: Sierra Club—Ballantine.

"Electricity: Benefits / Effects." 2004. 4 November <http://www.aboutnuclear.org/view.cgi?fC=Electricity,Benefits_^_Effects>.

Epstein, Richard A. 1985. *Takings: Private Property and the Power of Eminent Domain*. Cambridge, MA and London, England: Harvard University Press.

Epstein, Richard and Walter Block. 2004. Debate on Eminent Domain: http://maroon.uchicago.edu/news/articles/2004/05/04/block_epstein_will_d.php; 5/4/04; http://www.mises.org/blog/archives/002009.asp.

"EU Regulations: Fishermen 'Should Have Stake in Stocks'." December 15, 2005. *EIU ViewsWire*. New York.

Flew, A. 1987. "Social Science: Making Visible the Invisible Hands." *The Journal of Libertarian Studies*, Vol. 8, No. 2, pp. 197–212.

Flitner, Michael. 1998. *Privatizing Nature*. Edited by Michael Goldman. Great Britain: Pluto Press, pp. 144–167.

Flournoy, Alyson C. 2003. "In Search of an Environmental Ethic," *Colum. J. Envtl. L.*, Vol. 28, p. 63, 80.

Foreman, David. 1990. "Only Man's Presence Can Save Nature," *Harpers* (April).

Fornos, Werner. 1991A. "Gaining People, Losing Ground," *The Humanist*, May/June, pp. 5–6.

Fornos, Werner. 1991B. "The Environmental Crisis: A Humanist Call for Action," *The Humanist*, November/December, pp. 30–32.

Frey, B.S., W.W. Pommerehne, F. Schneider and G. Gilbert. 1984. "Consensus and Dissension Among Economists," *The American Economic Review*.

Friedman, David. 1973. *The Machinery of Freedom: Guide to a Radical Capitalism*. New York: Harper and Row.

Garrison, Roger W. 1985. "Inter-temporal Coordination and the Invisible Hand: An Austrian Perspective on the Keynesian Vision," *History of Political Economy*, Vol. 17, No. 2, pp. 309–321.

Gonzalez, Abel J. 1996. "Chernobyl–Ten Years After." *IAEA Bulletin*, Vol. 38, No. 3. http://www.dwb.unl.edu/Teacher/NSF/C03/C03Links/www.iaea.or.at/worldatom/inforesource/bulletin/bull383/gonzalez.html (last accessed January 20, 2006).

Goodman, John C. and Richard L. Stroup et al. 1991. *Progressive Environmentalism; A Pro-human, Pro- Science, Pro Free Enterprise Agenda for Change* (National Center for Policy Analysis, Task Force Report, Dallas, TX).

Gordon, David. 1990. *Resurrecting Marx: The Analytical Marxists on Freedom, Exploitation, and Justice*. New Brunswick, NJ: Transaction.

Gore, Al. 1992. *Earth in the Balance: Ecology and the Human Spirit*. New York: Plume.

Gore, Al. 2000 [1992]. *Earth in Balance*. New York: Houghton Mifflin Company, pp. 126–144.

Graber, David M. 1989. "Mother Nature as a Hothouse Flower," *Los Angeles Times Book Review* (22 October).

Grampp, W.S. 1950. "Some Effects of Rent Control," *Southern Economic Journal*, April, pp. 425–426.

Greenpeace. 2002a. "Can Genetic Engineering Feed the World?" A Greenpeace Report. http://archive.greenpeace.org/~usa/reports/biodiversity/ftw.html (accessed October 22, 2002).

Greenpeace Campaigns. 2002b. "Genetically-Engineered Food Claims Are Hard to Swallow" http://archive.greenpeace.org/~usa/campaigns/biodiveristy/factsheet/gefacts.html (accessed October 22, 2002).

Greenpeace. 2002c. "Plotting Behind the Fence." Tue 22 October. Belgium/Brussels (accessed October 22, 2002). www.greenpeace.org/international_en/news/details?new_id=47981.

Gwartney, James, Robert Lawson, and Walter Block. 1996. *Economic Freedom of the World, 1975-1995*. Vancouver: The Fraser Institute.

Hahn, Robert W. 1989. "Economic Prescriptions for Environmental Problems: How the Patient Followed the Doctor's Orders," *Journal of Economic Perspectives*, Vol. 3, No. 2, Spring, pp. 93–114.

Hahn, Robert W. and Gordon L. Hester. 1989. "Where Did All The Markets Go? An Analysis of EPA's Emissions Trading Program," *Yale Journal on Regulation*, Vol. 6, No. 1, Winter, pp. 109–133.

Hahn, Robert W. and Robert N. Stavins. 1990. "Incentive-Based Environmental Regulation: A New ERA from an Old Idea?," Harvard University Energy and Environmental Policy Center Discussion Paper, June 26.

Halbrooks, Jacob. 1997. *Value and the Environment*, Mises.org, Mar. 27, 2002, http://www.mises.org/story/922.

Hardin, Garret. 1989. "There is No Global Population Problem," in *The Humanist*, July/August, pp. 11–13, 32.

Hayek, Friedrich A. 1944. *The Road to Serfdom*. Chicago: University of Chicago Press.

Hayek, Friedrich A. 1945. "The Use of Knowledge in Society," *American Economic Review*, Vol. 35, No. 4, pp. 519–530.

Hayek, Friedrich A. 1948. "Socialist Calculation I, II, & III," *Individualism and Economic Order*. Chicago: University of Chicago Press.

Hayek, Friedrich A. 1973. *Law, Legislation and Liberty*. Chicago: The University of Chicago Press.

Hayek, Friedrich A. 1981. "The Repercussions of Rent Restrictions," *Rent Control: Myths and Realities*, edited by Walter Block. Vancouver: The Fraser Institute.

Hayek, Friedrich A. 1989. *The Fatal Conceit: The Errors of Socialism*. Chicago: The University of Chicago Press.

Heinlein, Robert. 1973. *Time Enough for Love*. New York: Berkeley Medallion, 1973.

Herbener, Jeffrey M. 1996. "Calculation and the Question of Mathematics," *Review of Austrian Economics*, Vol. 9, No. 1, pp. 151–162.

Herrick, D. 2007. "Wal-Mart Drug Discounts Increase Competition and Lower Prices," available at: www.heartland.org (accessed May 2, 2007).

Historical Statistics of the United States: Colonial Times to 1970, part 2; Washington, DC: U.S. Department of Commerce, Bureau of the Census, pp. 912, 13.

Hodel, Donald Paul. 1987. United States Department of the Interior. Arctic National Wildlife Refuge, Alaska, Coastal Plain Resource Assessment: Report and Recommendation to the Congress of the United States Final Legislative Environmental Impact Statement, April.

Hoff, Trygve J.B. 1981. *Economic Calculation in a Socialist Society*. Indianapolis: Liberty Press.

Hoppe, Hans-Hermann. 1989. *A Theory of Socialism and Capitalism: Economics, Politics and Ethics*. Kluwer Academic Publishers, Boston.

Hoppe, Hans-Hermann. 1993. *The Economics and Ethics of Private Property: Studies in Political Economy and Philosophy*. Boston: Kluwer.

Hoppe, Hans-Hermann. 1996. "Socialism: A Property or Knowledge Problem?," *Review of Austrian Economics*, Vol. 9, No. 1, pp. 147–154.

Hoppe, Hans-Hermann. 2001. *Democracy - The God That Failed: The Economics and Politics of Monarchy, Democracy, and Natural Order*. Rutgers University, NJ: Transaction Publishers.

Hoppe, Hans-Hermann and Walter Block. 2002. "Property and Exploitation," *International Journal of Value-Based Management*, Vol. 15, No. 3, pp. 225–236; http://www.mises.org/etexts/propertyexploitation.pdf.

Horowitz, David. 1991. *Deconstructing the Left*. Lanhan, MD: Second Thought Books.

Horwitz, Morton J. 1977. *The Transformation of American Law: 1780-1860*. Cambridge: Harvard University Press.

Horwitz, Steven. 1996. "Money, Money Prices, and the Socialist Calculation Debates," *Advances in Austrian Economics*, Vol. 3, pp. 59–77.

Hsieh, Chih-hao, et al. October 19, 2006. "Fishing Elevates Variability in the Abundance of Exploited Species," *Nature*, Vol. 443, pp. 859–862.

Hudson, Richard L. 1990. "Cost of Chernobyl Nuclear Disaster Soars in New Study---1986 Reactor Accident Dwarfs Other Soviet Peacetime Catastrophes," *The Wall Street Journal* (eastern edition). New York, NY. March 29, p. A8.

Johnson, Bruce. 1981. "The Environmental Costs of Bureaucratic Governance: Theory and Cases," in *Baden and Stroup's Bureaucracy vs. Environment*. The University of Michigan, pp. 217–221.

Johnson, Bruce, ed. 1982. *Resolving the Housing Crisis: Government Policy, Decontrol, and the Public Interest*. San Francisco: The Pacific Institute.

Jones, Gary L. "Testimony before the Subcommittee on Clean Air, Wetlands, Private Property, and Nuclear Safety, Committee on Environment and Public Works, U.S. Senate." Performance and Accountability Series: Major Management Challenges and Program Risks: Nuclear Regulatory Commission (GAO/OCG-99-19, Jan. 1999). Available online at: http://epw.senate.gov/107th/gao_2-4.htm (last accessed January 19, 2006).

Kahneman, Daniel, Jack L. Knetsch and Richard Thaler. 1986. "Fairness as a Constraint on Profit Seeking: Entitlements in the Market," *American Environment Review*, Vol. 76, No. 4.

Kaufman, Wallace. 1994. *No Turning Back: Dismantling the Phantasies of Environmental Thinking*. New York: Basic Books.

Keizer, William. 1987. "Two Forgotten Articles by Ludwig von Mises on the Rationality of Socialist Economic Calculation," *Review of Austrian Economics*, Vol. 1, pp. 109–122.

Keizer, Willem. 1997. "Schumpeter's Walrasian Stand in the Socialist Calculation Debate," in *Austrian Economics in Debate*, edited by Willem Keizer, Bert Riben and Rudy Van Zijp. London: Routledge.

Kinsella, Stephan N. 1996. "New Rationalist Directions in Libertarian Rights Theory," *Journal of Libertarian Studies*, Vol. 12, No. 2, Fall, pp. 313–326.

Kirzner, Israel M. 1973. *Competition and Entrepreneurship*. Chicago: University of Chicago Press.

Kirzner, Israel M. 1985. *Discovery and the Capitalist Process*. Chicago: University of Chicago Press.

Kirzner, Israel M. 1988. "The Economic Calculation Debate: Lessons for Austrians," *Review of Austrian Economics*, Vol. 2, pp. 1–18.

Klein, Peter G. 1996. "Economic Calculation and the Limits of Organization," *Review of Austrian Economics*, Vol. 9, No. 2, pp. 3–28.

Lange, Oscar and Fred M. Taylor. 1938. *On the Economic Theory of Socialism*. Minneapolis: University of Minnesota Press.

Lavoie, Don. 1981. "A Critique of the Standard Account of the Socialist Calculation Debate," *The Journal of Libertarian Studies*, Vol. V, No. 1, Winter, pp. 41–88.

Lavoie, Don. 1985. *Rivalry and Central Planning: The Socialist Calculation Debate Reconsidered*. New York: Cambridge University Press.

Lee, Dwight R. 2001. "To Drill or Not to Drill: Let the Environmentalists Decide," *The Independent Review*, Vol. VI, No. 2, Fall, Issue Number 1086-1653; pp. 217–226.

Lewin, Peter. 1998. "The Firm, Money and Economic Calculation," *American Journal of Economics and Sociology*, October.

Libecap, Gary D., and Ronald N. Johnson. 1979. "Property Rights, Nineteenth-Century Federal limber Policy and the Conservation Movement," *Journal of Economic Liberty,* pp. 129–142.

Lieberman, Bruce. October 20, 2006. "Fish Populations at Risk, Study Warns." *Copley News Service*. http://web.lexis-nexis.com (accessed February 02, 2007).

Lindbeck, Assar. 1972. *The Political Economy of the New Left*. New York: Harper and Row; cited in Rydenfelt, Sven, "The Rise, Fall and Revival of Swedish Rent Control," in *Rent Control: Myths and Realities*, Walter Block and Edgar Olsen, eds., 1981. Vancouver: The Fraser Institute, 1981, p. 213.

Lochbaum, David. 2000. *Nuclear Plant Risk Studies: Failing the Grade. Union of Concerned Scientists*. Cambridge: UCS Publications. Also available online at: http://www.ucsusa.org/energy.

Locke, John. 1948. "An Essay Concerning the True Origin, Extent and End of Civil Government," in *Social Contract*, edited by E. Barker. New York: Oxford University Press, pp. 17–18.

Locke, John. 1967. "An Essay Concerning the True Origin, Extent, and End of Civil Government," in *Two Treatises of Government*, edited by P. Laslett, 2nd ed. Cambridge: Cambridge University Press.

Lomborg, Bjorn. 2004. *The Skeptical Environmentalist: Measuring the Real State of the World.* Rev. ed. Cambridge University Press.

Lutins, Allen. 2019. "U.S. Nuclear Accidents." August 12; http://www.lutins.org/nukes.html.

Machan, Tibor. 1990. *Capitalism and Individualism: Reframing the Argument for the Free Society*. New York: St. Martin's Press, pp. 154–167.

Machan Tibor R. 1984. "Pollution and Political Theory," in *Earthbound: New Introductory Essays in Environmental Ethics*, edited by Tom Regan, Random House: New York.

Malek, N. 2002. "Myths about Business," www.mises.org (accessed April 29, 2007).

Mandel, Charles. 2006. "Fish Pirates Plundering Sea, Expert Says: Regulatory Agencies too 'Clumsy' to Stop Damaging Operations." *The Ottawa Citizen*. March 17, p. A4.

Markels, Jr., Michael. 1995. "Fishing for Markets: Regulation and Ocean Farming." *Regulation*, Vol. 18, No. 3, pp. 73–79. Washington, DC: Cato Institute.

Maser, Chris. 1996. *Resolving Environmental Conflict: Towards Sustainable Community Development*. Delray Beach, FL: St. Lucie Press, pp. 33, 146, and 178.

Mayell, Hillary. October 22, 2001. *National Geographic News*. "Controlled Alligator Harvest an Effective Conservation Tool, Louisiana Says". http://news.nationalgeographic.com/news/2001/10/1022_Ally_1.html. Site visited 11/5/06.

McChesney, F. 1991. "Antitrust and Regulation: Chicago's Contradictory Views," *Cato Journal*, Vol. 10, No. 3, pp. 775–798.

McGee, Robert W. and Walter Block. 1994. "Pollution Trading Permits as a Form of Market Socialism and the Search for a Real Market Solution to Environmental Pollution," *Fordham University Law and Environmental Journal*, Vol. VI, No. 1, Fall, pp. 51–77.

Meiners, Roger E. and Yandle, Bruce, eds. 1993. *Taking the Environment Seriously*. Contributors: Anderson, Terry L.; Benjamin, Daniel K.; Dalton, Brett A.; Heyne,

Paul; Leal Donald R.; Meiners, Roger E.; Nelson, Robert H.; Riggs, David W.; Shaw, Jane S.; Stroup, Richard L.; and Yandle, Bruce. Lanham, MD: Rowman and Littlefield Publishers, Inc, pp. 1–59, 224.

Metz, William C. 1983. "Legal Constraints to Repository Siting," in *Nuclear Waste: Socioeconomic Dimensions of Long-Term Storage*. Edited by Steve H. Murdock. Boulder: Westview Press, pp. 73–89.

Mills, Stephanie. 1989. *Whatever Happened to Ecology?* San Francisco: Sierra Club Books.

Mises, Ludwig von. 1975 [1933]. "Economic Calculation in the Socialist Commonwealth," in *Collectivist Economic Planning*, edited by F.A. Hayek. Clifton, NJ: Kelley.

Mises, Ludwig von. 1981 [1969]. *Socialism*. Indianapolis: Liberty Fund.

Morone, Joseph G. and Edward J. Woodhouse. 1989. *The Demise of Nuclear Energy? Lessons for Democratic Control of Technology*. New Haven: Yale.

Murphy, Arthur W. 1961. "Tort Liability, Government Indemnification, and Insurance," Conference on Atomic Radiation and the Law. The University of Chicago Law School, pp. 10–20.

Myrdal, Gunnar. 1965. "Opening Address to the Council of International Building Research in Copenhagen"; cited in *Dagens Nyheter* (Swedish Newspaper), 25 August 1965, p. 12; cited in Rydenfelt, Sven, "The Rise, Fall and Revival of Swedish Rent Control," in *Rent Control: Myths and Realities*, Walter Block and Edgar Olsen, eds., 1981. Vancouver: The Fraser Institute, p. 224.

National Audubon Society, www.audubon.org.

National Conference of State Legislatures. 2004. "Energy and Electric Utilities: State Laws and Regulations: Price Gouging," available at: www.ncsl.org (accessed April 12, 2007).

National Review. 1996. "Malthus Does Cairo," October 24, p. 18. https://www.emerald.com/insight/content/doi/10.1108/eb018839/full/html?skipTracking=true (accessed March 25, 2020).

Nelson, Robert H. 1997. "Does 'Existence Value' Exist? Environmental Economics Encroaches on Religion," *The Independent Review*, Vol. I, No. 4, Spring, pp. 499–521.

Nivola, Pietro S. 2004. "The Political Economy of Nuclear Energy in the United States," The Brookings Institution. Policy Brief #138 (September). <http://brookings.edu/comm/policybriefs/pb138.pdf (last accessed January 19, 2006).

Nozick, Robert. 1974. *Anarchy, State and Utopia*. New York: Basic Books.

Nuclear Energy Institute. 2004a. "Nuclear Power Plant Regulation: Sharpening the Focus on Safety," September. http://www.nei.org/doc.asp?catnum=3&catid=1262&docid=&format=print (last accessed January 19, 2006).

Nuclear Regulatory Commission. 2004b. "Part 50—Domestic Licensing of Production and Utilization Facilities," October. 2 November 2004 http://www.nrc.gov/reactors/new-reactor-licensing.html.

Nuclear Regulatory Commission. 2004c. 2 November; http://www.nrc.gov/reading-rm/doc-collections/news/2004/04-048iii.html.

Nutrition Action Health Letter. 2001. "Genetically Engineered Foods: Are They Safe," Center for Science in the Public Interest. November. U.S. Edition (accessed October 22, 2002).

Orr, Lloyd D. 1981. "Social Cost, Incentive Structures and Environmental Policies," in *Baden and Stroup's Bureaucracy vs. Environment*. The University of Michigan, p. 50.

Pasour, Jr., E. C. 1983. "Land-Use Planning: Implications of the Economic Calculation Debate," *The Journal of Libertarian Studies*, Vol. 7, No. 1, Spring, pp. 127–139.

Pigou, Arthur. 1912. *Wealth and Welfare*. London: Macmillan.

Pollak, Robert A. 1998. "Imagined Risks and Cost-Benefit Analysis." *The American Economic Review*, Vol. 88, No. 2, May, pp. 376–380.

Porrit, Jonathan and Winner, David. 1988. *The Coming of The Greens*. Fontana, London.

Porter, Michael E. and Claas van der Linde. 1995. "Toward a New Conception of the Environment-Competitiveness Relationship." *The Journal of Economic Perspectives*, Vol. 9, No. 4, Autumn, 97–118.

Posner, Richard A. 2004. *Catastrophe: Risk and Response*. Oxford University Press.

Rathje, William L. 1989. "Rubbish!," *Atlantic Monthly* (December), 99–109.

Revkin, Andrew. 2002. "Panel Urges U.S. to Tighten Approval of Gene-Altered Crops." *The New York Times*. February 22. http://query.nytimes.com/search/article- page.html?res=9E00E4DF133EF931A15751C0A9649C8B63 (accessed November 4, 2002).

Reyes, Sonia. 2002. "Organic Rules!" Brandweek. October 21. Page 3. www.brandweek.com.

Ricardo, David. 1912/1821. *The Principles of Political Economy*, 3rd ed., London: J.M. Dent.

Riccio, Jim. 2001. "Risky Business: The Probability and Consequences of a Nuclear Accident," Greenpeace USA. Available online at: http://www.greenpeace.org/raw/content/usa/press/reports/risky-business-the-probabilit.pdf (last accessed January 19, 2006).

Richman, Sheldon L. 1981. "War Communism to NEP: The Road from Serfdom," *The Journal of Libertarian Studies*, Vol. V, No. 1, pp. 89–98.

Rifkin, Jeremy. 1980. *Entropy: A New World View*. New York: Bantam.

Rifkin, Jeremy. 1987. "Time Wars: A New Dimension Shaping Our Future," *Utne Readers* (September).

Rifkin, Jeremy. 2002. *The Hydrogen Economy: The Creation of the World-Wide Energy Web and the Redistribution of Power on Earth*. New York: Penguin Putnam Inc.

Robichaux, Mark. 1989. *The Wall Street Journal (Eastern edition)*. "Alligator Farming Shows There's a Lot to Be Said for Cows—But High Prices on Reptiles Make Risks Worthwhile for Louisiana Farmers," August 2, New York, p. 1.

Rockwell, Jr., Llewellyn H. and Walter E. Block. 2010. "The Economics and Ethics of Hurricane Katrina," *American Journal of Economics and Sociology*, Vol. 69, No. 4, October, pp. 1294–1320; http://www.walterblock.com/wp-content/uploads/V.335_The-Economics-and-Ethics-of-Hurricane-Katrina.pdf.

Rosenthal, D.H. and R.H. Nelson. 1992. "Why Existence Value Should Not Be Used in Cost-benefit Analysis," *Journal of Policy Analysis and Management*, Vol. 11, No. 1, pp. 116–122; http://www.osti.gov/energycitations/product.biblio.jsp?osti_id=6657978.

Rothbard, Murray N. 1970. *Man, Economy and State*. Los Angeles, CA: Nash.

Rothbard, Murray N. 1971a. "Nixon's NEP," *The Libertarian Forum*, Vol. 3, No. 10.

Rothbard, Murray N. 1971b. "Lange, Mises and Praxeology: The Retreat from Marxism." *Toward Liberty*. Menlo Park, CA: Institute for Humane Studies, Vol. II, pp. 307–321. Reprinted in *The Logic of Action One: Method, Money, and the Austrian School*. Glos, UK: Edward Elgar Publishing Ltd., 1997, pp. 384–396.

Rothbard, Murray N. 1973. *For a New Liberty*. New York: Macmillan.

Rothbard, Murray N. 1976. "Ludwig von Mises and Economic Calculation Under Socialism," in *The Economics of Ludwig von Mises*, edited by Laurence S. Moss. Kansas City: Sheed & Ward, pp. 67–77.

Rothbard, Murray N. 1982a. "Law, Property Rights, and Air Pollution," *Cato Journal*, Vol. 2, No. 1, Spring; reprinted in *Economics and the Environment: A Reconciliation*, Walter Block, ed., Vancouver: The Fraser Institute, 1990, pp. 233–279; http://www.mises.org/rothbard/lawproperty.pdf.

Rothbard, Murray N. 1982b. *The Ethics of Liberty*. Atlantic Highlands, NJ: Humanities Press.

Rothbard, Murray N. 1986. "Government vs. Natural Resources,'" *The Free Market*, pp. 1–5.

Rothbard, Murray N. 1990. "Law, Property Rights, and Air Pollution," in *Economics and the Environment: A Reconciliation*, edited by Walter Block. Vancouver: The Fraser Institute, pp. 233–279; http://www.mises.org/rothbard/lawproperty.pdf.

Rothbard, Murray N. 1991. "The End of Socialism and the Calculation Debate Revisited," *Review of Austrian Economics*, Vol. 5, No.2, pp. 51–70.

Rothbard, Murray N. 1995. "Vouchers: What Went Wrong," in *Making Economic Sense*. Auburn, AL: Mises Institute; ch. 43; http://www.mises.org/econsense/ch43.asp.

Rothbard, Murray N. 1997 [1956]. "Toward a Reconstruction of Utility and Welfare Economics." Reprinted in "The Logic of Action" Vol. I. Lyme, NH: Edward Elgar. pp. 225–227.

Rubin, Charles T. 1994. *The Green Crusade: Re-Thinking the Roots of Environmentalism*. New York: McMillan Free Press.

Runolfsson, Birgir. 1997. "Fencing the Oceans: A Rights-Based Approach to Privatizing Fisheries." *Regulation*, Vol. 20, No. 3, pp. 57–62. Washington, DC: Cato Institute.

Saign, Geoffrey. 1994. *Green Essentials: What You Need to Know About the Environment*. San Francisco: Mercury House, pp. 30–43.

Sale, Kirkpatrick. 1989. "Presidential Matters," *Resurgence* 132 (January–February).

Salerno, Joseph. 1990. "Postscript: Why A Socialist Economy is 'Impossible,'" *Ludwig von Mises, Economic Calculation in the Socialist Commonwealth*. Auburn, AL: Mises Institute, pp. 51–71.

Salerno, Joseph. 1995. "A Final Word: Calculation, Knowledge and Appraisement," *Review of Austrian Economics*, Vol. 9, No. 1, pp. 141–142; http://www.mises.org/journals/rae/pdf/rae9_1_8.pdf.

Salins, Peter D. 1980. *The Ecology of Housing Destruction: Economic Effects of Public Intervention in the Housing Market*. New York: New York University Press.

Salzman, Lorna. 1979. "The World's Most Dangerous Garbage," in *Accidents Will Happen: The Case Against Nuclear Power*. New York: Harper & Row, pp. 110–127.

Samuelson, Paul A. 1956. "Social Indifference Curves," *Quarterly Journal of Economics,* Vol. 70, No. 1, February, 1–22.

Samuelson, Paul A. 1970. *Economics*, 8th ed. New York: McGraw Hill.

Sanders, B. 2005. "Stop Oil Company Price Gouging," available at: www.common-dreams.org (accessed March 11, 2007).

Schumacher, F. F. 1973. *Small is Beautiful*. New York: Harper and Row.

Scriven, Tal. 1997. Wrongness, Wisdom, and Wilderness: Toward a Libertarian Theory of Ethics and the Environment. State University of New York Press.

Sherman and Jaffe of the Nutrition Action Healthletter. "Genetically Engineered Foods: Are They Safe?" http://ww.cspinet.org/nah11_01/ disclose.

Shogren, Elizabeth. 2002. "New Testing Policy Proposed for Altered Crops." *Los Angeles Times*. www.latimes.com/business/la-fi-food3auf03.story?null (accessed November 4, 2002).

Shugart, W.F. II.1987. "Don't Revise the Clayton Act, Scrap It!" *Cato Journal*, Vol. 6, pp. 925.

Siegan, Bernard H. 1970. "Non-Zoning in Houston," *Journal of Law and Economics*, Vol. XIII, No. 1, April.

Siegan, Bernard. 1972. *Land Use Without Zoning*. Lexington, MA: Heath.

Simon, Julian. 1981. *The Ultimate Resource*. Princeton: Princeton University Press.

Smith, Adam. 1776/1965. *An Inquiry into the Nature and Causes of the Wealth of Nations*. New York: Modern Library.

Smith, Andy. 1995. "Woman of Color and Productive Choice: Combating the Population Paradigm," *Journal of Feminist Studies in Religion*, Vol. II, Fall, pp. 39–66.

Smith, Craig. 2002 "China Rushes to Adopt Genetically Modified Crops." *The New York Times*. Travel. http://query.nytimes.com/search/article page.html?res=9 D0CE0D6153CF934A35753C1A9669C8B63 (accessed November 4, 2002).

Smith, Fred L. Jr. 1983. "Why Not Abolish Antitrust?" *Regulation*, January–February, 23, pp. 23.

Snyder, Pamela and Jane Shaw. 1995. "PC Oil Drilling in a Wildlife Refuge," *Wall Street Journal,* September, 7.

Sowell, Thomas. 1983. *The Economics and Politics of Race*. New York: Morrow.

Sowell, Thomas. Undated. http://www.jewishworldreview.com/cols/sowell081800.asp

Spooner, Lysander. 1966 [1870], *No Treason*. Colorado: Larkspur.

State of New York Office of the State Comptroller, Division of Management Audit. 1996. "Long Island Power Authority Staff Study: Disposition of the Shoreham Nuclear Power Plant," Report 95-D-38. February 21. Available at: http://www.osc.state.ny.us/audits/audits/9596/95d38.pdf (last accessed January 19, 2006).

Steele, David Ramsay. 1981. "Posing the Problem: The Impossibility of Economic Calculation Under Socialism," *The Journal of Libertarian Studies*, Vol. V, No. 1, Winter, pp. 7–22.

Steele, David Ramsay. 1992. *From Marx to Mises: Post Capitalist Society and the Challenge of Economic Calculation*, La Salle, IL: Open Court.

Sternglass, Ernest J. 1981. *Secret Fallout: Low-level Radiation from Hiroshima to Three Mile Island.* McGraw-Hill. Available online at http://www.ki4u.com/Secret_Fallout/SF.html (last accessed January 19, 2006).

Stigler, George. 1961. "The Economics of Information," *Journal of Political Economy,* Vol. 69, pp. 213–225.

Strasma, Jan and Viktoria Mitlyng. 2004. "NRC Seeks Public Input on Environmental Impact Statement for Proposed D.C. Cook Nuclear Plant License Renewal," Nuclear Regulatory Commission, October, 28.

Stroup, Richard L. 1988. "Buying Misery with Federal Land," *Public Choice,* Vol. 57, pp. 69–77.

Stroup, Richard L., et al. 1991. Progressive Environmentalism: A Pro-Human, Pro-Science, Pro-Free Enterprise Agenda for Change (Nat'l Ctr. for Pol'y Analysis, Dallas, Tex.).

Stroup, Richard L. 2003. *Economics: What Everyone Should Know about Economics and the Environment*. Washington, DC: The Cato Institute.

Tucker, William. 1990. *The Excluded Americans: Homelessness and Housing Policies*. Washington, DC: Regnery Gateway.

United States Fish and Wildlife Service. 2002. Arctic National Wildlife Refuge. "Purposes of the Arctic National Wildlife Refuge." October 2. http://www.r7.fws.gov/nwr/arctic/purposes.html.

Vance, Laurence. 2004. "Book Review of Clint Bolick. *Voucher Wars: Waging the Legal Battle over School Choice*. Washington, D.C.: Cato Institute, 2003." *Journal of Libertarian Studies,* Vol. 18, No. 2, Spring, pp. 99–102.

Walker, Michael. 1980. *Privatization: Theory and Practice*. Vancouver: The Fraser Institute.

Walker, Michael. 1988. *Privatization: Tactics and Techniques*. Vancouver: The Fraser.

Waters, Richard. 2006. *Financial Times*. "City Watcher Trials Use of ID Chips in Staff IT Hardware; [London 1St Edition]". February 13, London: pg. 28.

Whitehead, Roy and Walter Block. 2002. "Environmental Takings of Private Water Rights: The Case for Full Water Privatization," *Environmental Law Reporter*, October, pp. 11162–11176.

Whitehead, Roy and Walter Block. 2000. "Environmental Justice Risks in the Petroleum Industry," 24 *WM. & Mary Envtl. L. & Pol'y Rev.* 67.

Whitehead, Roy, Catherine Gould, and Walter Block. 2004. "The Value of Private Water Rights: From a Legal and Economic Perspective," *Albany Law Environmental Outlook Journal*, Vol. 9, No. 2, pp. 333–343. www.cspinet.org/nah/11_01/ (accessed October 22, 2002).

Yandle, Bruce. 2000. "Public Choice and the Environment: From the Frying Pan to the Fire," in *Terry Anderson's Political Environmentalism: Going Behind the Green Curtain*. Stanford, CA: Hoover Institutional Press, Stanford University.

Credits

Here is the original source of chapters that comprise this book; the author and publisher of this book thank the first publishers of these articles as well as my coauthors, for their permission to allow these publications to appear, once again, in this volume. Thanks to Leith Edgar for editorial assistance.

SECTION I. INTRODUCTION

Chapter 1. Private Property Rights and Environmentalism

Block, Walter. 1998. "Environmentalism and Freedom: The Case for Private Property Rights," *Journal of Business Ethics*, Vol. 17, No. 6, December, pp. 1887–1899.

SECTION II. WATER

Chapter 2. Ocean Real Estate

Motichek Serigny, Amy, Walter Block and Jay Johnson. 2008. "Forget Ocean Front Property, We Want Ocean Real Estate!" *Ethics, Place, and Environment*; Vol. 11, Issue 2, June, pp. 147–155. (Amy Motichek Serigny graduated from Loyola University New Orleans).

SECTION III. ENERGY

Chapter 3. Nuclear Power

Morrell, Joseph, Walter Block and John Levendis. 2006. "Nuclear Power." *Journal of Business Ethics*; Vol. 67, No. 1, August, pp. 37–49; http://dx.doi.org/10.1007/s105

51-006-9003-y. (Joseph Morrell is a graduate of, and John Levendis is a professor of economics at, Loyola University New Orleans).

Chapter 4. Energy Petition Memo to the NTU

Block, Walter. 2007. "Memo to the NTU on its energy petition" August 22; http://www.lewrockwell.com/block/block83.html.

Chapter 5. Oil and Gas Stations

5A. Block, Walter. 1991. "Gas Station Food Sales and Bicycle Rickshaws," *Fraser Forum,* June, pp. 42–44, http://oldfraser.lexi.net/publications/forum/1991/june/FF-06-91.html#gas. (first published in *British Columbia Report*, May 6, 1991; reprinted by Fraser Forum with permission).

5B. Block, Walter. 1990. "Beaufort Oil," *Fraser Forum,* September, p. 27. (I am grateful to Kristen McCahon of the Fraser Institute for helping me access material published under their auspices).

SECTION IV. PROPERTY RIGHTS

Chapter 6. Homesteading

Block, Walter. 1990. "Earning Happiness Through Homesteading Unowned Land: a comment on 'Buying Misery with Federal Land' by Richard Stroup," *Journal of Social, Political and Economic Studies*, Vol. 15, No. 2, Summer, pp. 237–253; http://www.walterblock.com/publications/happiness_homesteading.pdf.

Chapter 7. Property in Space

Huebert, Jacob. and Walter Block. 2007. "Space Environmentalism, Property Rights, and the Law." *Memphis Law Review*, Vol. 37, No. 2, Winter, pp. 281–309 (Jacob Huebert earned his J.D. degree from the University of Chicago).

Chapter 8. Land for Parks

8A. Block, Walter. 8/2/01. "Hunt Farms," http://www.lewrockwell.com/orig/block3.html.

8B. Block, Walter. 1992. "Stanley Park Task Force," *Fraser Forum*, October, pp. 40–42.

8C. Block, Walter. 7/17/89. "How much parkland?" *Fraser Forum*, August, pp. 23–24.

8D. Block, Walter. 1977. "Public Parks: the New York City Case." *The Libertarian Forum*. Vol. 10, No. 10, October, pp. 6–7; http://www.mises.org/journals/lf/1977/1977_10.pdf. (*The Libertarian Forum* is in the Public Domain).

SECTION V. ENVIRONMENTAL ECONOMICS

Chapter 9. The Economics of the Arctic National Wildlife Refuge

Schaeffer, Emily and Walter Block. 2007. "The Economics of the Arctic National Wildlife Refuge: A Casualty of Leasing Law or Candidate for Privatization." *Energy and Environment.* Vol. 18, No. 1, pp. 75–85 (Emily Schaeffer earned a BS at Loyola University New Orleans and Ph.D. degree at George Mason University).

Chapter 10. Genetically Modified Foods

10A. Christie Laporte, and Walter Block. 2005. "A Free-Market Environmentalist Approach to Genetically Modified Foods." *Santa Clara Journal of International Law;* Vol. III, Issue, No. 2. http://www.scu.edu/scjil/current/index.shtml; http://www.scu.edu/scjil/archive/v3_BlockAbstract.shtml; http://www.scu.edu/scjil/archive/v3_BlockArticle.shtml (Christie Laporte earned a BS and JD at Loyola University New Orleans).

10B. "Margarine." This article was first published in *British Columbia Report,* December 31, 1990, and was reprinted by *Fraser Forum* with permission.

Chapter 11. Economics of the Environment

11A. Block, Walter. 1991. "The Handicapped and the Environment," *Fraser Forum,* January, pp. 21–23; originally appeared here: Block, Walter. 1990. "The Handicapped and the Environment," *British Columbia Report,* v. 2, n. 12, November 19, pp. 26.

11B. Block, Walter. 1990. "The Gaia Hypothesis and Sustainable Growth," *Fraser Forum,* July, pp. 28–29; originally appeared here: Block, Walter. 1990. "The Gaia Hypothesis and Sustainable Growth," *British Columbia Report,* v. 1, n. 37, May 21, pp. 16.

11C. Block, Walter. 1990. "Sustainable Growth, Slow Growth," *Fraser Forum,* December, pp. 21–22; originally appeared here: Block, Walter. 7/30/90. "Slow growth, sustainable growth," *British Columbia Report.* p. 37.

11D. Block, Walter. 1991. "The Ivory Import Ban." *Regulation: The Cato Review of Business and Government.* Winter, pp. 27–28.

11E. Block, Walter. 1990. "A Plan To Save Our Endangered Species," *British Columbia Report,* v. 1, n. 31, April 9, p. 19.

11F. Block, Walter. 1990. "Peter Pearse: A Modern Day Henry George," *Fraser Forum,* November, pp. 28–29; this article was first printed in *B.C. Report* and was reprinted with permission.

11G. Block, Walter. 2000. "Of Swamps and Jungles," June 14; http://www.mises.org/fullstory.asp?control=446&FS=+Of+Swamps+and+Jungles+

11H. Block, Walter. 1991. "Environmental Gluttons," *Fraser Forum* June, pp. 32; http://oldfraser.lexi.net/publications/forum/1991/june/FF-06-91.html#gluttons;

11I. Block, Walter. 1990. "L'Affaire David Suzuki," *Fraser Forum*, March, pp. 23–24.

11J. Block, Walter. 1992. "Shooting Cats in a Cage," *Fraser Forum*, February, pp. 33–34; http://oldfraser.lexi.net/publications/forum/1992/february/FF-02-92.html #shooting.

11K. Block, Walter. 1990. "Sustainable Growth, Slow Growth," *Fraser Forum*, December, pp. 21–22; originally appeared here: Block, Walter. 7/30/90. "Slow growth, sustainable growth," *British Columbia Report*. p. 37.

11L. Block, Walter. 1991. "The Ivory Import Ban." Regulation: *The Cato Review of Business and Government*. Winter, pp. 27–28.

11M. Block, Walter. 1990. "A Plan To Save Our Endangered Species," *British Columbia Report*, v. 1, n. 31, April 9, p. 19.

11N. Block, Walter. 1990. "Peter Pearse: A Modern Day Henry George," *Fraser Forum*, November, pp. 28–29; this article was first printed in B.C. Report and was reprinted with permission.

Chapter 12. Pollution

12A. Block, Walter. 1994. "Pollution," *Cliches of Politics*, Mark Spengler, ed., Irvington on Hudson, New York: Foundation for Economic Education, pp. 267–270.

12B. Santoriello, Andrea and Walter Block. 1996. "Externalities and the Environment," *The Freeman*, November, Vol. 46, No. 11, pp. 755–756, http://www.fee.org/vnews .php?nid=3628; reprinted in El Diario, (Bolivia), July 1997 (Andrea Santoriello earned the BA at Holy Cross College).

12C. Block, Walter. 1990. "Pollution Monitoring," *Fraser Forum*, August, pp. 20.

12D. Block, Walter. 3/2/89. "Solving woes of solid wastes," *Daily Bulletin* tba;

12E. Block, Walter. 1989. "Recycling," *Fraser Forum*, March, pp. 27–28.

12F. Walter. 1989. "The Waste Makers." *Consent*. No. 10. Sep – Nov, pp. 1- 5tba

SECTION VI. CLIMATE; NATURAL DISASTERS

Chapter 13. New Orleans' Future

Block, Walter and Llewellyn H. Rockwell, Jr. 2007. "Katrina and the Future of New Orleans," Telos, Vol. 139, Summer, pp. 170–185; http://tinyurl.com/2wv8lc; http://journal.telospress.com; http://journal.telospress.com/cgi/reprint/2007/139/170.

Chapter 14. Weather Socialism

Block, Walter. 2006. "Katrina: Private Enterprise, the Dead Hand of the Past, and Weather Socialism; An Analysis in Economic Geography." *Ethics, Place and Environment: A Journal of Philosophy & Geography*; Vol. 9, No. 2, June, pp. 231–241; reprinted in 'Post-Katrina: Risk Assessment, Economic Analysis and Social Implications'– edited by Harry Richardson, Peter Gordon and James Moore. Edward Elgar Publishing.

Chapter 15. Katrina

Block, Walter. 2005. "Government and the Katrina Crisis." *The Free Market.* Vol. 26, No. 10, October; http://www.mises.org/freemarket_detail.asp?control=565&sortorder=articledate

15A. Block, Walter. 2004. "Me and hurricane Ivan." September 20; http://www.lewrockwell.com/block/block44.html;

15B. Block, Walter. 2005B. "The Answer to Katrina." September 11, http://www.lewrockwell.com/block/block53.html;

15C. Block, Walter. 2005. "Government and the Katrina Crisis." The Free Market. Vol. 26, No. 10, October; http://www.mises.org/freemarket_detail.asp?control=565&sortorder=articledate

Chapter 16. Price Gouging

Culpepper, Dreda and Walter Block. 2008. "Price Gouging in the Katrina Aftermath." *International Journal of Social Economics*; Vol. 35, No. 7, pp. 512–520;

SECTION VII. POPULATION

Chapter 17. Armageddon?

Coffey, Daniel and Walter Block. 1999. "Postponing Armageddon: Why Population Growth Isn't Out of Control," *Humanomics*. Vol. 15, No 4, pp. 66–79

Chapter 18. Cause of Poverty?

English, Patrick and Walter Block. 1997. "Overpopulation or Lack of Economic Freedom: Which Is the Real Cause of Poverty?" *Fundamentals of Prosperity*, Essay #2, February, Center for World Capitalism of the James Madison Institute, pp. 1–11.

Chapter 19. Problem?

Block, Walter. 1989. "Population Growth: Is It a Problem?" *Resolving Global Problems into the 21st century: How Can Science Help Proceedings of the Fourth National Conference of Canadian Pugwash*, Peter S. Ross, Sheila Riordon and Susan MacArtney, eds., Ottawa: CSP Publications, pp. 30–61.

Index

Note: Italicized pages refer to figures and tables; page numbers followed by "n" denote endnotes

About the Author

Walter E. Block is Harold E. Wirth endowed chair and professor of Economics, College of Business, Loyola University New Orleans, and senior fellow at the Mises Institute. He earned his PhD in economics at Columbia University in 1972. He has taught at Rutgers, SUNY Stony Brook, Baruch CUNY, Holy Cross, and the University of Central Arkansas. He is the author of more than 600 refereed articles in professional journals, two dozen books, and thousands of op-eds (including the *New York Times*, the *Wall Street Journal*, and numerous others). He lectures widely on college campuses, delivers seminars around the world, and appears regularly on television and radio shows. He is the Schlarbaum Laureate, Mises Institute, 2011; and has won the Loyola University Research Award (2005, 2008) and the Mises Institute's Rothbard Medal of Freedom, 2005; and the Dux Academicus award, Loyola University, 2007. Prof. Block counts among his friends Ron Paul and Murray Rothbard. He was converted to libertarianism by Ayn Rand. Block is old enough to have played chess with Friedrich Hayek and once met Ludwig von Mises, and shaken his hand. Block has never washed that hand since. So, if you shake his hand (it's pretty dirty, but what the heck) you channel Mises.

www.ingramcontent.com/pod-product-compliance
Ingram Content Group UK Ltd.
Pitfield, Milton Keynes, MK11 3LW, UK
UKHW020840290926
15470UKWH00004B/1452

9 781498 586856